United States
Department of
Agriculture

Natural
Resources
Conservation
Service

NATIONAL FORESTRY HANDBOOK

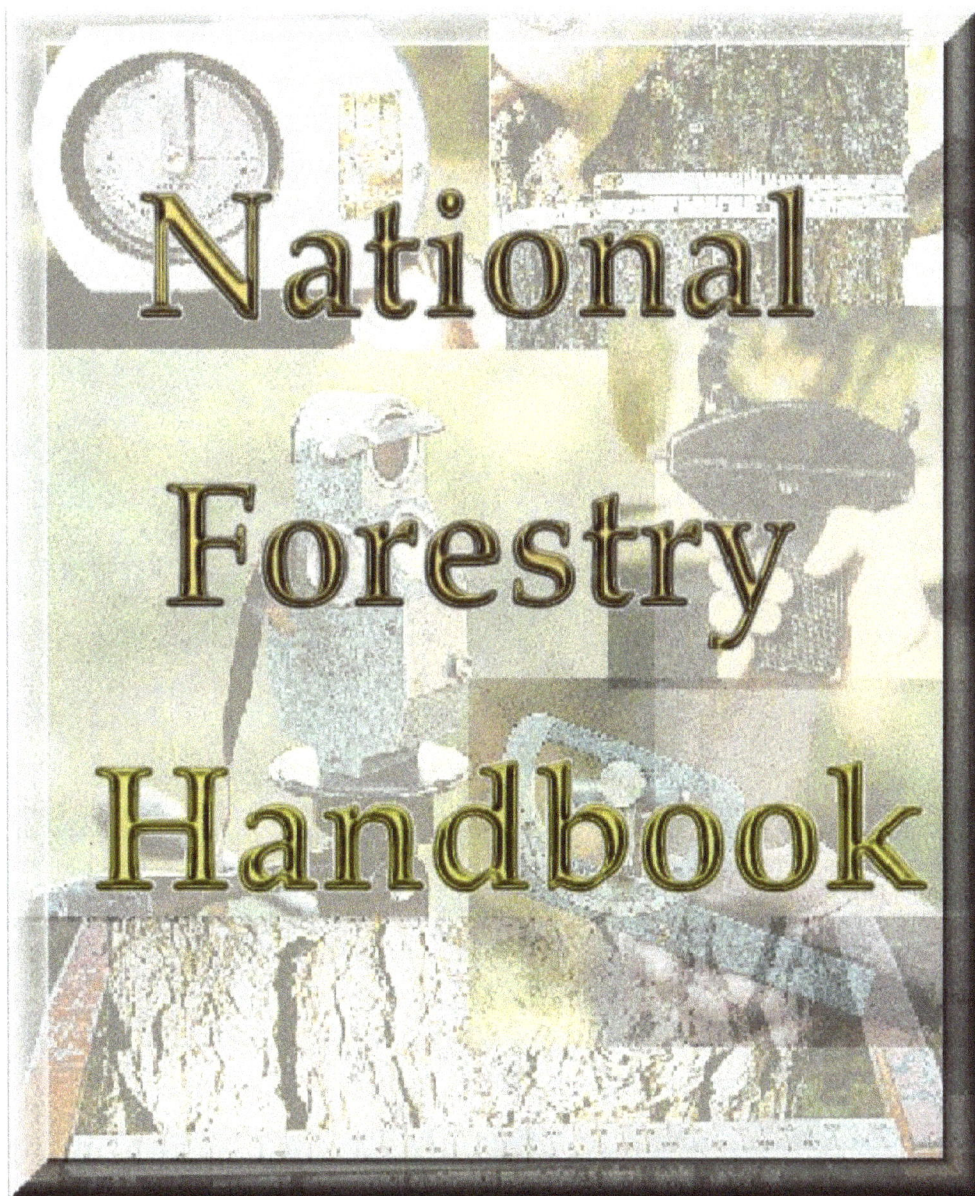

Title 190   NATIONAL FORESTRY HANDBOOK
Issued, February 2004

Recommended Citation:
Natural Resources Conservation Service, National Forestry Handbook, title 190, February 2004.

The United State Department of Agriculture (USDA) prohibits discrimination in its programs on the basis of race, color, national origin, sex, religion, age, disability, political beliefs and marital or familial status. (Not all prohibited bases apply to all programs.) Persons with disabilities who require alternative means for communication of program information (Braille, large print, audiotape, etc.) should contact USDA's TARGET Center at 202-720-2600 (Voice and TDD). To file a complaint, write the Secretary of Agriculture, U.S. Department of Agriculture, Washington, D.C.20250 or call 1-800-245-6340 (voice) or (202) 720-1127 (TDD). USDA is an equal employment opportunity employer.

United States
Department of
Agriculture

Natural
Resources
Conservation
Service

NATIONAL FORESTRY HANDBOOK

PREFACE

The National Forestry Handbook, as a subdivision of the NRCS directives system, includes parts 635 through 638.

The format is intended to allow flexibility for additions and updates.

The National Forestry Handbook (NFH) contains methodology, procedures, and related reference materials that assist NRCS personnel to implement the policies contained in the National Forestry Manual (NFM) relative to forestry and agroforestry technologies.

All references to the Soil Conservation Service or SCS by Public Laws, Memoranda or other documents stated herein have been changed to the Natural Resources Conservation Service or NRCS, respectively.

All policies and responsibilities relating to forestry previously assigned to the Soil Conservation Service are carried forward in full to the Natural Resources Conservation Service unless otherwise noted or amended in this handbook.

PART 635 – OPERATIONS AND MANAGEMENT

PART 636 – CONSERVATION PLANNING

PART 637 – SOIL-RELATED FORESTRY AND AGROFORESTRY INTERPRETATIONS

ix

PART 638 – INFORMATION SYSTEMS

TABLE OF FIGURES

PART 635 – OPERATIONS AND MANAGEMENT

CONTENTS
PART   **PAGE**

Part 635.0 – General

635.00 Mission and Objectives

The mission of NRCS is to provide leadership and administer programs to help people conserve, improve, and sustain our natural resources and environment. Toward this end, NRCS is committed to conservation forestry by maintaining a high level of expertise in planning, using and conserving soil, water, animals, plants, air, and related human resources.

635.01 Purpose

The purpose of the National Forestry Handbook (NFH) is to provide informational material to assist Natural Resource Conservation Service personnel in the planning and application of forestry and agroforestry practices on nonfederal forestland throughout the United States. The NFH contains methodology, procedures, and related reference materials. This reference material assists NRCS personnel in implementing NFM policy in forestry and agroforestry technologies.

635.02 Supplementing the Handbook

States are encouraged to supplement this handbook with material relative to the application of forestry and/or agroforestry in their locales. Supplements must be in accordance with the NRCS Directives System. Copies of all state-level supplements will be provided to the Director of Ecological Sciences Division at National Headquarters.

635.03 Relationship to the National Forestry Manual

The National Forestry Manual (NFM) is a companion document to the NFH. The NFH contains methodology, procedures, and related reference materials. This reference material assists NRCS personnel to implement NFM policy in forestry and agroforestry technologies. The NFH consists of Parts 635, 636, 637 and 638, which directly complement NFM Parts 535, 536, 537 and 538 respectively. Materials prepared for the NFH will be numbered based on the predominant relationship to specific paragraphs in the NFM. NFH material may be prepared at any administrative level. The originating level is responsible for administrative and technical support of such materials published in the NFH.

635.04 Quality Assurance

Refer to the National Forestry Manual, Part 537 for policy and guidance relative to quality assurance reviews.

635.10 General

Refer to the National Forestry Manual for the authorities that govern the forestry program activities of the Natural Resources Conservation Service

635.20 General

635.30 General

Refer to the National Forestry Manual for specific guidance and policies on cooperating with non-government organizations.

Refer to the National Forestry Manual for specific guidance and policies on cooperating with other agencies.

635.40 General

This section provides examples of state-developed technical materials, guidelines on how to develop technical notes, and potential sources of forestry-related training. Parts 635.41, 635.42, 635.43, and 635.45 are intentionally omitted to maintain coordination to numbering in the National Forestry Manual.

635.44 Technology Transfer

(a) Acquiring and Maintaining Technical Materials

It is the responsibility of each individual to keep their technical knowledge up-to-date. This section provides information on some of the many resources available to keep abreast of the current technology forestry and agroforestry activities.

Forest and range experiment stations (see U.S Forest Service, below) publish a considerable volume of periodic research papers and notes on forestry topics.

Many national publications contain research articles valuable for foresters. The Journal of Forestry, Forest Science, Ecology, Journal of Soil and Water Conservation, National Agriculture Library and many others can be obtained through circulation lists, library services, or public libraries.

Many experiment stations welcome group tours of the station, written requests to individual researchers, or telephone calls for information.

The following are potential sources that individuals may find useful in keeping their technical knowledge up to date.

(1) U.S. Forest Service (USFS)
The Forest Service conducts research through a network of forest and range experiment stations and the Forest Products Laboratory. Detailed information on specific topics of research can be found on each research station's Web site. Thousands of publications are available for either downloading or ordering through the research station. The location and Web site address of each research station regional site is listed below.

- International Institute of Tropical Forestry
  Río Piedras, Puerto Rico
  PO Box 25000, Río Piedras, PR 00928-5000
  http://www.fs.fed.us/global/iitf/welcome.html

- Pacific Southwest Research Station
  Box 245
  Berkeley, California 94701
  510-559-6300
  http://www.psw.fs.fed.us

- Pacific Northwest Research Station
  333 SW 1st Avenue, PO Box 3890
  Portland, OR 97208-3890503-808-2592
  http://www.fs.fed.us/pnw

- North Central Research Station
  1992 Folwell Avenue
  St. Paul, Minnesota 55108
  651-649-5173
  http://www.ncfes.umn.edu

- Northeastern Research Station
  11 Campus Boulevard Suite 100
  Newtown Square, PA 19073
  610-557-4017
  http://www.fs.fed.us/ne

- Southern Research Station
  Box 2680
  Asheville, North Carolina 28802
  http://www.srs.fs.fed.us

- Rocky Mountain Research Station
  240 West Prospect Road
  Fort Collins, Colorado 80526-2098
  970-498-1100
  http://www.fs.fed.us/rm

- Forest Products Laboratory
  One Gifford Pinchot Drive
  Madison, Wisconsin 53705
  608-231-9200
  http://www.fpl.fs.fed.us

(2) Cooperative State Research, Education, and Extension Service (CSREES)
CSREES links the research and education programs of the U.S. Department of Agriculture to provide access to scientific knowledge; strengthen the capabilities of land-grant and other institutions in research, extension and higher education; increase access to improved

communication and network systems; and promote informed decisionmaking. CSREES works closely with the following institutes to develop the research and education programs needed to solve agricultural, environmental and community problems.

- Land-grant institutions in each state, territory and the District of Columbia

- More than 130 colleges of agriculture; 59 agricultural experiment stations; 57 cooperative extension services

- 63 schools of forestry

- 17 1890 historically black land-grant institutions, including Tuskegee University

- 27 colleges of veterinary medicine

- 42 schools and colleges of human sciences

- 29 1994 Native American land-grant institutions

- 190 Hispanic-serving institutions

The forest product programs of the Cooperative State Research, Education, and Extension Service (CSREES) are conducted in cooperation with universities in every State, and with other agencies (Forest Service (FS), Natural Resources Conservation Service (NRCS) , and Agricultural Research Service (ARS) and organizations. For information about CSREES forest products research, visit the Forestry Programs Web site at: http://www.reeusda.gov/1700/programs/forestry.htm

(3) The Agricultural Research Service (ARS)
The Agricultural Research Service (ARS) is the principal research agency of the U.S. Department of Agriculture (USDA). It is one of the four component agencies of the Research, Education, and Economics (REE) mission area. ARS scientists communicate research results and transfer new technologies from the agency to other scientists, institutions of higher education, producers, product and process developers, consumers, and other end users through publications; conferences, workshops, and consultations; and cooperative agreements and patent licenses.

Research stations conduct wind erosion research to increase the understanding of wind erosion processes, develop reliable predictive tools and improve windbreak design technology. For information on current wind erosion research efforts, visit the Agricultural Research

Service Web site at: http://www.ars.usda.gov.

(4) Natural Resources Conservation Service (NRCS)
NRCS provides onsite assistance to conservation districts and their cooperators. The resources available include:

- Published soil surveys – http://soils.usda.gov

- Ecological site inventory information relative to forest productivity and windbreak performance – http://plants.usda.gov – ESI Link

- Ecological site descriptions – http://plants.usda.gov – ESD Link

- Soil-related forestry and agroforestry interpretations http://soils.usda.gov/technical/nfhandbook – Part 537

This above information is used in technical guides and soil survey reports that support activities conducted by field offices for conservation districts and their cooperators.

NRCS operates 26 Plant Materials Centers (PMC's) throughout the Nation. The PMC's assemble, evaluate, select, and distribute promising new plants for a wide range of conservation uses. To learn more about the activities at the various PMC's, visit their Web site at: http://plant-materials.nrcs.usda.gov/

(5) Colleges and Universities
Colleges and universities work with the Cooperative State Research, Education, and Extension Service (CSREES) in connection with the McIntire-Stennis Forestry Research Act. In addition, many staff members do part-time research funded by grants from other organizations and industrial firms. Other researchers are funded full-time from state appropriations. Several have forest laboratories in addition to campus facilities. Close working relationships are maintained with forest and range experiment stations that have laboratories at state colleges and universities. For a list of accredited foresty institutions visit the Society of American Foresters Web site at: http://www.safnet.org/educate/pforschools.htm.

(6) Forest Industry
Most of the larger forest industry companies maintain laboratories and teams of scientists to carry out investigative work important to their needs. Much of this is centered around forest products research, particularly in pulp and paper, particle board, and plywood, and in

635-5

forest management research for raising, tending, and harvesting the tree crops. Associations of forest industries also specialize in research on some related products. Manufacturers of forest harvesting equipment do limited research specific to the use of their products. A good resource for finding Internet information on the forest products industry, forest products, wood science and technology, wood engineering, forest products marketing, wood industry associations, wood industry consultants, wood industry data sources, and wood-products-based research can be found at http://www.forestdirectory.com

(b) Disseminating Technical Information

The National Forestry Manual gives policy and guidance for issuing technical information at the area, state, or national level. One effective method of disseminating technical information is with the use of Technical Notes. The following are some helpful tips on writing effective Technical Notes. Several examples of effective Technical Notes can be found on the National Agroforestry Center's Web site at: http://www.unl.edu/nac/

- **Narrow the topic**
  Write on a specific topic rather than a bigger or more general subject
- **Simplify the content**
  Avoid jargon and big words
  Make technical information easy to understand
  Avoid unnecessary phrases and adjectives
  Use short sentences
- **Be concise**
  Include only necessary information
- **Break up blocks of text**
  Use short sections with titles, subtitles, and bullets
- **Answer these questions**
  Who? What? When? Where? Why? and especially, How?
- **Organize in sections**
  Introduction – What is it?
  Application – Where does it apply?
  Benefits
  How to design it
  How to apply it: Step 1, Step 2, etc.
  Follow-up and maintenance needs
- **Know your audience**
  Natural resource professionals? Landowners?
  Write on what is important to them

- **Include additional resource information**
  Publications
  Where to get help (e.g. local NRCS office, state forestry, etc.)

(c) Training

Foresters and others involved in forestry and agroforestry activities need to take advantage of available training opportunities in order to maintain technical competency. The following is a list of some of the many training resources available to NRCS personnel. In addition to those resources listed below, training is also available at numerous colleges and universities throughout the country and from state and/or regional NRCS offices.

- The National Employee Development Center (NEDC) is the national training division of the NRCS. NEDC provides educational and developmental activities through traditional and alternative delivery systems that meet the NRCS's mission and goals and enables employees to achieve individual and career goals. Visit their Web site at http://www.ncg.nrcs.usda.gov/nedc/homepage.html for more information.

- The NRCS offers many in-house technical training courses throughout the country. A description and contact person for these courses can be found at http://www.ftw.nrcs.usda.gov/pdf/Train.pdf.

- The U.S. Fish & Wildlife Service's National Conservation Training Center (NCTC) in West Virginia provides training and education services to conservation professionals from a variety of agencies and organizations. Visit their Web site at http://www.nctc.fws.gov for more information.

- The Forest Service Continuing Education (CE-WFRP) Program is designed to meet the training needs of entry-level and mid-career professionals. In addition to Forest Service personnel, a number of openings are available for resource specialists from other Federal agencies. Visit their Web site at http://www.fs.fed.us/biology/education for more information.

- The USDA Graduate School offers career-related courses to all adults regardless of education or place of employment. The Graduate School annually provides more than 1,500 different courses for career development and personal enrichment. Classes are designed to help individuals realize their career potential and improve their job performance. Visit their Web site at http://grad.usda.gov for more information.

- The U.S. Geological Survey Biological Resources Division, through the Midcontinent Ecological Science Center (MESC) offers a special selection of natural resources management courses. Instructed courses are scheduled throughout the year at various locations. Others are available as correspondence courses and independent study. The courses are open to individuals from federal, state, and local governments; universities; private businesses; and foreign governments. Visit their Web site at http://www.mesc.usgs.gov/training/mesc-training.html for more information.

- The NRCS Graduate Studies Program is a unique opportunity for selected employees to enter a formalized graduate study program at an accredited college or university.

  The NRCS Graduate Studies Program will pay tuition and related expenses to selected employees for the purpose of enhancing the agency's expertise in targeted priority developmental needs areas. The program is open to employees at the GS-11 grade level and above (or equivalent pay systems). The

program is offered for full-time and part-time study. Curriculum is offered in the technical and management field. Visit their Web site at http://www.ncg.nrcs.usda.gov/nedc/grad_grant.html for more information.

- The National Arbor Day Foundation has a wide range of programs that provide opportunities for continuing education for professionals in government services. Visit their Web site at http://arborday.org for more information for more information.

- The Society of American Foresters (SAF) offers a Continuing Forestry Education Program designed to reward and recognize those who pursue a program of continuing education and professional development. The program is available nationwide and is open to members and nonmembers of SAF. Visit their Web site at http://www.safnet.org/educate/index.html for more information.

CONTENTS

636.00 Introduction

General Manual 180–CPA, Part 409 establishes Natural Resources Conservation Service (NRCS) policy that guides NRCS employees as they assist clients in planning and implementing resource conservation plans.

The NRCS National Planning Procedures Handbook provides guidance on the "how to" of the planning process as related to the planning policy established by the General Manual.

The National Forestry Handbook (NFH) provides the "how to" of forestland resource conservation planning. This handbook provides guidance and information concerning the planning process, specifically for forestry and agroforestry related activities. The NFH provides the technical guidance for developing resource information for inclusion in the Field Office Technical Guide (FOTG).

General Manual 450–TCH, Amendment 4, Part 401 establishes NRCS FOTG policy. The local FOTG contains the technical information needed to assist clients in the development and application of conservation plans. It contains general resource information about the field office area, soil and site information (forestland ecological sites, conservation tree/shrub groups, rangeland ecological sites, and forage suitability groups), quality criteria to be met by Resource Management Systems (RMS's), guidance documents depicting the resource management planning thought process, practice standards for all practices applicable to the local field office area, and examples of the Conservation Effect Decision Making Process.

The VegSpec decision support system provides automated assistance in working with forestry and agroforestry clients to develop their conservation plans.

636.01 Objectives

The objectives of forestry and agroforestry related conservation planning on forestland are to assist clients to

- Understand the basic ecological principles associated with managing their land – the soil, water, air, plants, and animals.

- Realize they are part of a complex ecosystem and that their management decisions influence the ecological changes that occur.

- Realize their responsibilities and the importance of protecting the environment and maintaining future options for the use of the resource.

- Develop and implement a plan that meets the needs of the soil, water, air, plant, and animal resources and their management objectives.

- Recognize the productivity potential of their forest resource for wood, forage, and other nontimber-related forest products.

636-v

Conservation plans include decisions for manipulating the plant community to manage the soil, water, air, plant, and animal resources. These five resources are clearly related and respond to each other in an interactive mode. On forestland, plants are the resource that directly affects the soil, water, air, and animal resources.

The major objective in forestry- and agroforestry-related conservation planning is the design and establishment of forestry and agroforestry practices that, when coupled with any necessary facilitating practices, will meet the quality criteria for the five resources established in the local FOTG and also the objectives of the client. When properly implemented, these conservation plans benefit the client, the local community, and the Nation. Well-managed forestlands, along with the carbon sink they afford, the clean water and air they produce, the recreation they provide, and the plants and animals they support, make a major contribution to the natural beauty of the landscape and to the maintenance of a quality and economically sound environment.

636.02 Planning Procedures

Refer to Parts 636.1 through 636.4 for detailed forestry and agroforestry planning procedures.

636.10 General

Success in the application of forestry and agroforestry practices is directly related to the quality of the assistance provided to the client during the planning process.

The National Planning Procedures Handbook (NPPH) describes the NRCS planning process. The NPPH details how the planning process is applied in developing viable and meaningful plans for managing, protecting, and restoring natural resources. Planners need to be thoroughly familiar with the guidance and principles described in the NPPH.

NRCS assists clients who own or control the land for which conservation plans are being prepared. It must be understood that

- Clients make the decisions.
- Clients apply the practices and pay for them.
- NRCS is assisting them in preparing their plans.
- Conservation planning is productive only when firm decisions have been made by the client.

Recording practices in a conservation plan by NRCS personnel, when the client has not made the decision to apply the plan, is not appropriate planning that leads to the application of resource management systems. Conservation planning is productive when clients understand their ecosystem to the degree that their daily decision-making is influenced and their decisions in the conservation plan reflect this stimulus.

636.11 Preplanning

The National Planning Procedures Handbook (NPPH), Part 600.16 describes the items that need to be addressed before planning activities begin. Preplanning is of major importance to the effectiveness of the planning process. Preplanning includes the following activities:

(a) Gather Materials and Information

- *Site Interpretation Information*
  Gather interpretive information needed for the conservation planning process, such as:
  Forestland ecological site descriptions
  Rangeland ecological site descriptions
  Conservation tree/shrub groups

  Use and management interpretations
  Forestland productivity interpretations
  Plant identification materials; a good source is the Plants Web site at http://plants.usda.gov

- *Conservation Practice Standards*
  Understand all the conservation practices applicable to forestry and agroforestry activities in Section IV of the Field Office Technical Guide.

- *Inventory Methods and Tools*
  Prepare for the field inventory. Have at hand and thoroughly understand the use of the various methods and tools needed to inventory the planning area's resources. See Parts 636.2 and 636.3 for a detailed description of the common inventory methods and tools used in forestry and agroforestry applications.

- *Soil Information*
  Forest planning begins with a soil survey. Planners need to be familiar with soils of the area and the forestry/agroforestry interpretations that apply. Forestry/agroforestry interpretations describe the behavior and limitations of soil components as they relate to vegetation, productivity, and management. See the National Forestry Manual, Part 537 for detailed documentation of forestry and agroforestry interpretations.

  Study the soil survey information for the area under consideration and consult the field office technical guide (FOTG) for forestry and/or agroforestry interpretive information relative to the appropriate soil components. The FOTG contains forestry and agroforestry interpretive information relative to soil map units such as: potential forest productivity; use and management limitations or hazards; kinds of trees likely to be present; adapted species for forestry and/or agroforestry practices; and which species are most suited for wood production, agroforestry practices (block, row, or density plantings), wildlife, aesthetic values, etc.

- *Aerial Photographs*
  Study the aerial photograph of the area under consideration. Use a stereoscope to locate drainageways, streams, ridges, developed areas, and other features that might affect the use of the land. Locate logging roads and other disturbed areas on the photo that need to be visited for planning necessary erosion control measures. When planning

636-3

agroforestry practices, such as windbreak or shelterbelt establishment, locate important features such as buildings, crop boundaries, roads, feedlots, etc.

Outline, on the map, those areas that need onsite examination. Apparent differences in cover, changes in aspect or soil, and natural barriers are used to make tentative field separations. Forested areas are not necessarily "fenced in," as are cropland and pasture. A boundary of a forest is usually determined by a change of soil or forest condition or by a natural barrier.

If available, GIS technology can be utilized to provide much of the above information.

- *Applicable Laws, Regulations, and BMP's*
  Be familiar with and understand all of the national and local laws and regulations that may have an impact on conservation planning activities. This would include state "Best Management Practices," where applicable.

(b) Prepare Yourself

- Be knowledgeable about the basic ecological principles of forestland in your work area and be prepared to discuss them in a manner that land managers can understand.

- Be able to determine forest productivity, forest health, plant community dynamics, trees to manage, and adapted species.

- Understand the quality criteria for soil, water, air, plants, and animals as specified in Section III of the FOTG.

- Understand and be proficient in the nine steps of conservation planning.

- Understand and be proficient in the use of the VegSpec decision support system to assist in the planning process.

- Be knowledgeable about the cost of implementing forestry and agroforestry-related practices (tree planting, timber stand improvement, conservation tree/shrub establishment, etc.).

- Be knowledgeable about the economics (cost-return) relative to forestry and agroforestry activities (logging, hunting, Christmas trees, silvopasture, etc.)

- Be knowledgeable of the cultural and social resources applicable to the planning area.

- Be knowledgeable of any threatened or endangered species that may be indigenous to the planning area.

(c) Determine Client's Objectives

Determine as much as possible about the client's desires, objectives, and level of knowledge of ecological principles on forestland. Secure this information from notes in current conservation plans and by visiting with other personnel who may have worked with the individuals on prior occasions.

(d) Make Firm Planning Schedule

Make firm dates with the clients and discuss the purpose of the planning dates. Ensure that they understand the time requirements in order to schedule sufficient time for the planning dates. Arrive at the assigned time, prepared for the day's work.

(e) Ensure Understanding of Basic Ecological Principles

Ensure that the client has a basic knowledge of ecological principles. Important items to know and understand are:

- Identity of plants on their land
- Effects of forestry and agroforestry management techniques (fire, harvesting, thinning, timber stand improvement, and other management decisions that affect the ecological dynamics of the plant community)
- How plants compete with each other in plant communities
- Ecological site concepts (explain the soil-plant relationship)
- Selection of trees and shrubs for agroforestry applications
- Forest understory reactions to canopy manipulation
- Forest production and habitat values of the different plant communities that can exist on a forestland site
- Multiple use opportunities
- How forestland is managed to protect or improve water quality
- Wildlife needs for food, water, and cover

636-4

An understanding of these basic principles by clients is essential to the planning process. Without this knowledge they cannot continuously inventory and analyze their resources, recognize problems and their cause, develop proper and obtainable objectives, formulate and evaluate treatment alternatives, plan a course of action, implement the plan, and continuously evaluate results and make improvements.

636.12 Identify the Problem

(a) General

When clients contact NRCS requesting assistance, they have perceived a problem and want to solve it. The perceived problem may actually be a symptom caused by the real problem. An example: the client has recognized an invasion of undesirable vegetation in their forest stand and an increased incidence of disease in the desirable trees. To the client, these are definite problems, but both are symptoms of the problem of poor timber management. The lack of applying timber stand improvement techniques (prescribed fire, chemical deadening) has allowed undesirable plants to become established and reproduce. This has led to increased competition for water and nutrients within the forest stand. The increased competition and the lack of properly timed harvesting of the desirable trees have resulted in a stress condition in which the desirable trees are more susceptible to disease and insect infestation.

The problem was not what the client originally perceived, but rather the lack of sound timber management that caused the symptoms. The NRCS objective is to help land managers recognize real problems, not just symptoms. When poor timber management is a problem, the NRCS conservationist should not tell managers the problem is poor timber management. Instead, **the conservationist must lead them to recognize that poor timber management is the problem.** This can be accomplished by helping them understand their ecosystems as described in preplanning. The process of recognizing the problem continues from preplanning through the steps of resource inventory and analyzing the resource data.

(b) Standard

Land managers are led to recognize the symptoms and causes of problems through an understanding of the ecosystem and the inventory process.

(c) Activities

The activities needed to identify the problem are shown below.

What	How
Clients identify perceived problems	Personal observations, often without the knowledge required to identify the cause of the problem.
Clients develop an understanding of ecosystems	NRCS personnel ensure that land managers understand their ecosystems by teaching and showing them on their land.
Clients recognize the real problems and the causes of problems	NRCS assists land managers to • Inventory the resources in the ecosystem. • Identify all the symptoms—soil, air, plant, and animal problems and potential problems—and the causes of each. • Recognize all the causes of symptoms as resource problems that must be addressed in the planning process.

636.13 Determine the Objectives

(a) General

All clients have a set of objectives. These objectives may or may not include the proper management of the ecosystem to accomplish their desired objective. If not, the reason may be a lack of understanding of all the interactions in the ecosystem. To assist clients in the planning process, objectives must be established by them after they fully understand the ecosystem, have inventoried the resources, and identified the problems. When working with clients, it is often best not to ask for firm objectives until these three processes have been accomplished. Some people do not like to change their minds once they have made a firm commitment to an objective. Assist them to understand and inventory their resources and identify the problems before they express their objectives.

(b) Standard

NRCS employee leads the client to develop ecologically and economically sound objectives.

(c) Activities

The activities needed to determine the objectives are shown below.

What	How
Client expresses management objectives.	This is often accomplished without a sound understanding of ecological principles, resource inventories, or problems identified.
Client expresses objectives for management that are ecologically, economically, and socially sound.	NRCS personnel • Ensure that client understands the ecosystem. • Assist managers in inventorying their resources. • Assist managers in recognizing resource problems and causes. • Assist clients to establish objectives that are ecologically, economically, and socially sound.

636.14 Inventory the Resources

(a) General

Once clients understand the ecological principles of their land, they generally ask:
"What is the productivity potential of my forestlands?"
"What is the forest health?"
"How does my forest stand compare to its potential?"

At this point the clients are beginning to understand the dynamics of the ecosystem and the fact that it is important to determine and understand as much as possible about their lands. They will desire your assistance in inventorying the resources.

(b) Standard

NRCS employees assist clients in inventorying their ecosystems and facilitating practices currently in place (current harvesting schemes, current timber stand improvement practices, forest stand productivity, wildlife habitat and numbers, etc.). During this process the conservationist should develop an understanding of the client's available resources to implement the conservation plan.

(c) Activities

The activities needed to inventory the resources are shown below. Refer to Parts 636.2 and 636.3 for details on inventory methods and tools.

What	How
Secure needed materials for inventory	NRCS secures maps (aerial photos and soil maps), equipment used in the field, and technical information, such as range ecological site descriptions, forestland ecological site descriptions, forage suitability groups, and conservation tree/shrub groups.
Conduct the inventory	NRCS personnel • Assist the client to identify forestland and rangeland ecological sites, conservation tree/shrub groups, and forage suitability groups on aerial photos from soil interpretations and ground truthing. • Determine site indices and record on the plan map. • Determine forest stand distribution and indicate on map. • Record land boundaries, roads, logging trails, farmsteads, and other important features on the plan map. • Complete wildlife habitat evaluations. • Determine soil erosion, condition, and contamination. • Identify sediment depositions. • Evaluate water quality and water yield. • Determine wildlife numbers and condition. • Develop forage inventory. • Develop livestock and wildlife inventory. • Develop forage and animal needs balance sheet. • Identify active and potential recreation resources and client's aesthetic considerations. • Identify cultural resources, if present. • Identify endangered plant and animal species and habitat, if present. • Identify available resources.

636.15 Analyze Resource Data

(a) General

After the inventory process is complete, an analysis of the data is necessary to assist the client to identify and quantify problems. Again, it is imperative for clients to understand ecosystem concepts before they can analyze resource data. Only then can they understand the relationship of soil, water, air, plant, and animal resources in ecosystems and the causes of resource problems. The analysis may point out opportunities that the client has not recognized, such as

- Recreation – Fee hunting, camping, bed and breakfast, fishing, hiking, and bird-watching.
- Alley Cropping – Planting of trees or shrubs in conjunction with agronomic, horticultural, or forage crops.
- Forest Farming – Cultivating high-value specialty products in conjunction with growing timber, such as medicinals and herbs, decorative items, specialty wood items, edibles, etc.
- Silvopasture – Managing timber and livestock as a single integrated system.

(b) Standard

NRCS assists client in analyzing the inventory data so they may recognize resource problems, as well as new opportunities.

(c) Activities

The activities needed to analyze resource data are shown below.

What	How
Evaluate the current ecosystem in relation to site potentials	NRCS assists land managers to determine: • Whether the current plant community provides the desired attributes of production, habitat, water quality and quantity, air quality, and soil protection. • Desirability of tree species for the products, production, and economic returns desired. • Health of existing forest stands. • Forage value rating for livestock. • Effects of current management programs on the plant and animal communities. • Wildlife habitat values in relation to potential for the site. • Significance of cultural resources, if present. • Endangered or threatened plant or animal species, if present. • Opportunity for new enterprises.

636.16.1 Formulate Alternative Solutions

(a) General

Phase II of the planning process begins with development of alternative solutions. At least one of the alternatives developed should be a Resource Management System (RMS), meeting the quality criteria for all resource problems identified and the objectives of the client. The Conservation Effects Decision (CED) worksheets can be used to present the impact of the RMS and other alternatives to the client. In developing Resource Management System alternatives, vegetation management practices will be planned that meet the needs of the plants and animals. Facilitating practices (log landings, haul roads, tree planting, windbreaks, etc.) will be planned when needed to enable the application of the planned vegetative management practices. Facilitating practices (diversion terraces, water bars, stabilization structures, etc.) will be planned when needed to treat specific problems.

(b) Standard

NRCS employees will assist the client to develop treatment alternatives that meet quality criteria in the FOTG for resource problems identified and that accomplish objectives of the client.

(c) Activities

The activities needed to identify the problem are shown below.

What	How
Develop treatment alternatives	Select the vegetation management and facilitating practices that will meet quality criteria established in the local FOTG for all resource problems identified, and that meet the management objectives of client. Develop sufficient numbers of alternatives from which client's may select an alternative to meet their needs.

636.16.2 Evaluate Alternative Solutions

(a) General

After alternative solutions are developed, client's evaluate them to determine which one best meets their objectives and solves the identified resource problems.

(b) Standard

Effects of each alternative are evaluated individually and compared to benchmarks for their ability to solve or alleviate identified resource problems and meet the client's objectives.

(c) Activities

The activities needed to evaluate alternative solutions are shown below.

What	How
Determine ecological, economical, and social effectiveness of treatment alternatives	Determine • Effectiveness of the alternative to achieve the desired plant community. • Effectiveness of each alternative to solve or alleviate each of the soil, water, air, plant, and animal resource problems. • Economic and social feasibility of each alternative. Decision support systems can assist in the economic evaluation of the treatment alternative. See Part 538 of the National Forestry Manual. • If the producer has the willingness, values, skills, and commitment to apply the system of practices.

636.17 Make Decisions

(a) General

After all the alternatives have been evaluated, the client makes a decision on which alternative meets their objectives. This is accomplished by comparing the alternatives to determine which one
- Will best achieve the desired plant community
- Will meet the desired time schedule
- Is the most financially and economically sound
- Is consistent with the client's knowledge and skills
- Is consistent with the client's time and distance restraints

After the alternative is selected, it is recorded in a manner that will assist the land manager in application.

Application of the selected alternative is usually a logical sequence that should be reflected in the schedule of application in the plan narrative. The following logic provides ideas for scheduling application.

If a forest stand is overstocked, then a thinning operation should be scheduled and applied. Facilitating practices such as logging roads, log landings, etc. must be installed before applying the thinning operation.

Maintenance practices should be scheduled when they are most advantageous. For instance, it may be advantageous to schedule prescribed burning ahead of the thinning operation to facilitate access to the area for marking timber. On the other hand, mechanical removal of large undesirable trees may be more advantageous after the thinning operation to prevent damage to harvestable timber from falling trees.

Facilitating practices (such as critical area treatment) would be applied last if harvest operations or prescribed burning would be detrimental to the newly established critical area vegetation. Each land unit will have its unique set of circumstances that dictate the schedule of application.

(b) Standard

NRCS leads the client to select alternatives that best meet the client's objectives. Decisions are recorded in the conservation plan.

(c) Activities

The activities needed to make decisions are shown on the following page.

636-11

What	How
Client selects best alternatives to meet client's objectives	NRCS assists the client in comparing each of the alternative evaluations to determine the one that best meets the objectives.
Schedule of application	NRCS personnel assist the client in developing a long-term schedule of application that ensures proper sequence and timing of applications for success.
Conservation plan prepared	NRCS assists the client in preparing the conservation plan. Client's copy should contain • Soil and water conservation district agreements. • Conservation plan maps, which should delineate as scale of map permits: – Operating unit boundary – Planned field boundaries, number, and acres – Land use of each field – Location of present and planned enduring practices – Ecological site delineation – Forage value ratings on grazed forest land – Other pertinent information, such as roads, log landing areas, riparian buffers, and wildlife corridors • Soils map and legend • All inventory data • Record of treatment alternatives selected and schedule of application • Fact sheets and/or job sheets NRCS case file contains • All information placed in the client's copy • Directions for location of the land unit(s) • List of job sheets furnished to the client • Technical assistance notes • Record of accomplishments

636.18 Implement the Plan

(a) General

The client is now ready to implement the plan. NRCS personnel shall provide technical assistance to the client in the application of all practices as needed and requested. For the plan to be successful, the client often requires close and continuous technical assistance from NRCS personnel as they learn to make the needed plant community observations and adjustments in management strategy.

For any planned practice or combination of practices, the technical guide gives the best known specifications. NRCS personnel must be familiar with and understand the specifications and other important materials in the technical guide. The standards and specifications include practice modifications that can be used to eliminate, overcome, or counteract any moderate or severe soil limitation shown in Section II of the technical guide. If the landowner is being assisted in the application phase by a State-employed forester or a consulting forester, the forester should be given the interpretations and urged to consider appropriate measures.

(b) Standard

NRCS provides technical assistance to the client to ensure the successful application of the planned practices.

(c) Activities

The activities needed to implement the plan are shown below.

What	How
Application of planned practices	NRCS personnel provide technical assistance to clients in the design and application of conservation plans. NRCS personnel must provide on-site assistance in a timely manner to continually teach clients to observe their resources and make the management decisions that will ensure success.
	Facilitating practices, such as logging roads and log landings all need to be installed according to a technical design to ensure success. NRCS personnel shall provide the on-ground technical assistance needed for design and installation.
	Facilitating practices, such as timber stand improvement, riparian buffers, forest site preparation, tree planting, prescribed burning, critical area treatment, diversions, streambank and shoreline protection, and structures for water control, all need to be installed according to a technical design to ensure success. NRCS shall provide the technical assistance needed for design and installation.

636.19 Evaluation of Results

(a) General

After clients initiate the application of their conservation plans, NRCS should provide follow-up assistance. Management is an ongoing process. The client may need the assistance of NRCS personnel to evaluate the results of the applied practices, such as timber harvesting or timber stand improvement. The initial planning process is just the beginning of learning and understanding of ecological management for many clients.

What	How
Provide needed follow-up for the evaluation of results, fine tuning of the management plan, revision of plan, and obtaining response data	Make firm dates with the client for follow-up evaluation assistance. Explain the purpose of the contact so the client may prepare. Review on the ground the results of the applied management practices. Use the opportunity to teach and assist clients to recognize trends in plant community response. Assist them to adjust their management to cause the plant community to respond as desired and meet the needs of the soil, water, air, plant, and animal resources. Review the schedule of application of facilitating practices. Review those that have been applied to evaluate their continued success. Assist in improving the schedule of application. Assist in recognizing any maintenance needed on applied practices. Gather response data that will improve the ability to predict future responses to treatment. Assist clients to identify new resource problems that need attention. Provide clients with new technical information applicable to their resource problems. Assist the clients to revise their conservation plans as needed. Follow the nine steps of conservation planning to accomplish this process.

and tools used in forestry and agroforestry applications. This part details the various methods used to conduct resource inventories related to the planning process.

Much of the understanding a client acquires about the nature of their resources, on which they may base many of their decisions, comes during the inventory stage. It is essential to work on the land with the decisionmaker

Part 573.5 for a detailed discussion of Forestland Ecological Site Descriptions.

636.21 Forest Stand Inventory

The zigzag transect method of sampling will normally meet the inventory needs relative to conservation planning. However, there may occasionally be a need to employ other common sampling methods.

In addition to the zigzag transect method, three other common forest stand inventory methods are also discussed in this subpart – strip sampling, fixed plot sampling, and variable plot sampling.

The strip sampling and plot sampling methods are based on a percentage system. A limited proportion of the area is measured, on the assumption that the samples are typical of the entire stand. The percentage of the area sampled depends on the uniformity of the stand and the size of the area to be sampled. In uniform stands, typical sampling percentages range from 20 percent on small areas of from 20 to 40 acres to 5 percent on areas larger than 80 acres. In areas where trees are of irregular distribution, the percentage of the area sampled may need to be increased to give adequate results.

Only a brief explanation of the strip sampling and plot sampling methods is given. Foresters should refer to other sources, such as the Society of American Foresters' *Forestry Handbook*, for a detailed description of these inventory methods.

(a) Zigzag Transect Method

A common inventory procedure used by NRCS foresters is the zigzag transect. The zigzag transect is a simple and rapid forest land inventory system that is used to determine

- Average tree diameter
- Range of tree diameters
- Stocking rates (trees per acre)
- Stand composition
- Stand condition (health)

(1) Zigzag Transect Procedures
The following procedures are used to conduct a zigzag transect.
(i) Step 1 – Select Main Stand
The main stand is usually made up of larger trees. There may be more than one general crown level. Beneath the main stand there is usually an understory of suppressed trees, advanced reproduction, or other plants. The client's principal concern should be with the main stand. (see Figure 636-1).

(ii) Step 2 – Choose a Route
Choose a route through the stand so you can sample a good cross section. Generally, this can best be accomplished by crossing the drainageways. On a sunny day you can use the sun as a direction marker by going toward it, away from it, or at some angle to or from it. A visible landmark can also be used as a direction marker.

(iii) Step 3 – Select a Starter Tree
The starter tree may be any tree that is a part of the main stand. No measurements are made of the starter tree. It serves only as a point of beginning.

Figure 636-1   **Differentiating Stands**

636-15

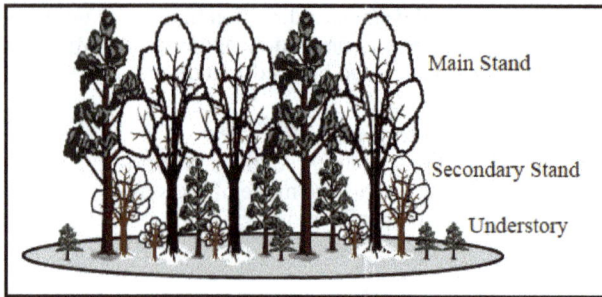

An occasional tree may be borderline between the main stand and the secondary stand. If, in your opinion, the tree offers significant competition to the tree in the main stand, consider it as part of the main stand.

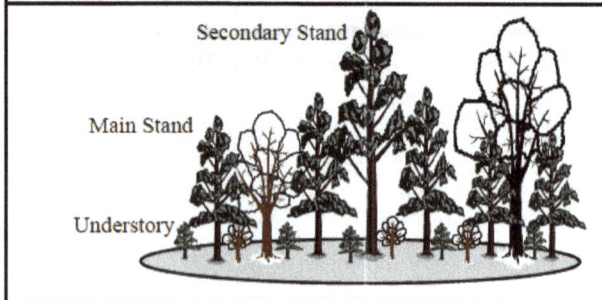

Don't separate large trees as a secondary stand unless they are considerably larger and clearly of an earlier generation than the trees of the main stand.

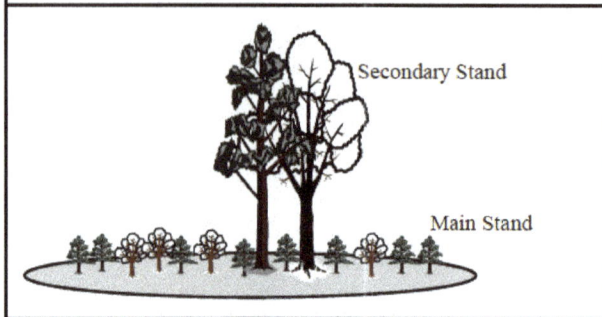

If the larger trees are numerous, there may be a question as to which is the main stand. In case of doubt, consider the larger trees as the main stand.

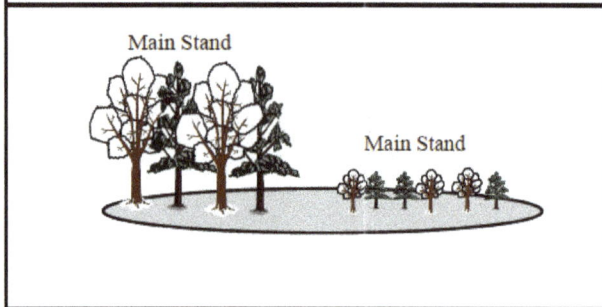

A change in the main stand may show need for a field boundary

(iv) Step 4 – Choose a Direction

At the base of the starter tree, face the chosen direction, place your heels together and position your toes to make a 90-degree angle. A line along the direction of travel bisects the angle formed by your feet (see Figure 636-2). A 90-degree arc is printed on some information sticks to help define the angle. When a 25" stick is held horizontally l2" from the eye, the ends of the stick form a 90-degree angle. A compass may also be used.

(v) Step 5 – Locate Closest Tree

Locate the closest main stand tree, the center of which is within the 90-degree angle. This is tree #1, as shown in Figure 636-3.

(vi) Step 6 – Determine Distance, Species, and Diameter

Pace or measure the distance from the center of the starter tree to the center of tree # 1. Determine the species of the tree identified in step 5 and measure its diameter at breast height (4.5 feet). Record measurements in the field notes (see Figure 636-5).

Figure 636-2   Starter Tree

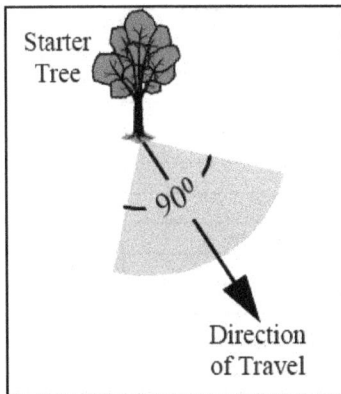

(vii) Step 7 – Rate Tree Condition

Examine the tree and rate its condition as good, fair, or poor. A good tree is reasonably straight, has a sound and full crown, does not have excessive limbs, and does not have evidence of scars, wounds, or disease. A poor tree may have a broken top, a bad crotch, excessive limbs, canker, wounds, scars, disease, or a combination of defects. Use fair as an intermediate rating. Do not confuse species desirability with the condition rating. Rate each tree on its merits, without regard to species.

Figure 636-3   Tree Selection Sequence

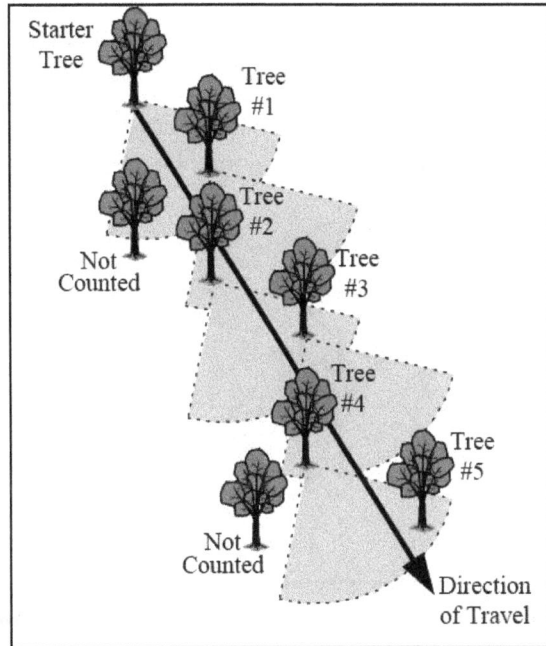

Record the condition rating in the field notes as shown in Figure 636-5. Show in the "notes" the reason for rating a tree as fair or poor.

(viii) Step 8 – Repeat Process

Standing at tree # 1, repeat steps 5-7 to select, measure, and rate tree # 2. Continue in this manner until at least 20 trees have been examined. The line of travel will proceed in a zigzag fashion as shown in Figure 636-3.

(2) Zigzag Conventions

The following conventions are observed when conducting a zigzag transect.

(i) Clumps and Open Area

Skip over openings and clumps or patches of trees that are not a part of the main stand or are decidedly different in kind or size from the main stand. Do not include spacing measurements or diameter measurements of trees on the edges of openings or clumps. Bypass those trees in the chosen direction of travel and start measurements on the opposite side (see Figure 636-4).

636-17

Figure 636-4   Openings and Clumps

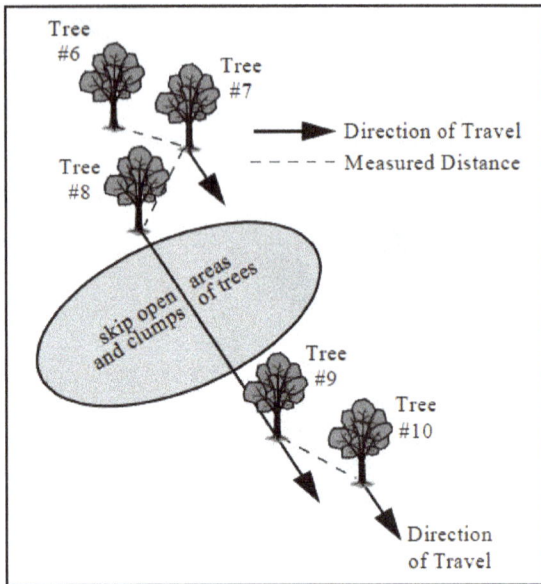

(ii) Plantations
In plantations, alternate the direction of travel. Use the direction of the row for the first tree; go at 90⁰ to the row for the second; use the direction of the row for the third, and so on.

(iii) Eligible Trees
When two eligible trees are equidistant, select the one closest to the direction of travel. Trees joined at the base are considered separate and individual and both may be counted.

(3) Zigzag Inventory Analysis
(i) Stand Diameter Calculation
Average stand diameter is obtained by dividing the total of diameters by the number of trees sampled. For example: Average diameter = 176 ÷ 20 = 8.8 inches; round to 9 inches.

The range of diameters can be determined by noting the smallest and largest of the trees sampled. (In the example, 6 to 11 inches).

Figure 636-5   Zigzag Transect Field Notes Example

Tree No.	Species	Distance (feet)	Diameter (inches)	Condition	Notes
1	loblolly pine	12	8	good	
2	loblolly pine	9	7	good	
3	loblolly pine	15	10	fair	scar at base
4	loblolly pine	16	11	good	
5	shortleaf pine	14	11	good	
6	shortleaf pine	5	9	good	
7	loblolly pine	13	8	poor	broken top
8	loblolly pine	14	8	good	
9	loblolly pine	9	9	good	
10	loblolly pine	14	6	good	
11	loblolly pine	9	7	fair	cronartium cankers
12	loblolly pine	10	7	good	
13	loblolly pine	11	9	good	
14	shortleaf pine	11	11	good	
15	loblolly pine	11	11	good	
16	loblolly pine	1	8	good	
17	loblolly pine	1	8	good	
18	loblolly pine	9	9	good	
19	loblolly pine	10	8	poor	cronartium cankers
20	shortleaf pine	15	11	good	
TOTALS		238	176		
AVERAGE		11.9	8.8		

(ii) Average Tree Spacing Calculation
Average tree spacing is found by dividing the total of distances by the number of trees sampled. For example: Spacing = 238 ÷ 20 = 11.9 feet; round to 12 feet.

(iii) Trees Per Acre Calculation
The number of trees per acre is calculated as follows:

$$\text{Number of trees per acre} = \frac{43560}{spacing^2}$$

For example: $\dfrac{43560}{12^2} = 303$ trees/acre

(iv) Thinning Determinations
For planning purposes, the $D + x$ "rule of thumb" is adequate to approximate the number of trees that need to be removed from a stand to avoid overcrowding. This rule of thumb is primarily applicable to even-aged stands. According to the $D + x$ rule, the average spacing between trees should equal the average stand diameter (D) plus a constant (x), expressed in feet. The constant x varies, depending on location and tree type. In Southern States, a constant of 6 is most commonly used for southern pines. In Western States, the constant can range from 2 for West Coast Douglas fir, to 4 for ponderosa pine. For stands with average diameters less than 6 inches, constants of 4 in the East and 2 in the West are commonly used. You should consult your local forest specialist to determine the constants applicable to the trees in your locale.

The approximate number of trees to be removed in a $D + x$ thinning is the difference between the number now present and the number that would be present after thinning. For example:

Assume that 6 is the applicable spacing constant for this stand of trees. From the zigzag transect, it is determined that the average tree diameter of the stand is 9 inches, and the average number of trees per acre is calculated to be 303. According the $D + x$ rule, the average trees per acre for 9 inch trees is calculated to be 194, as follows:

$D + 6 = 15$
$43560/15^2 = 194$

Therefore, approximately 109 trees per acre (303-194) need to be removed to provide adequate spacing.

See Exhibit 636-1 to determine the appropriated number of trees per acres at various D + spacings and tree diameters.

(v) Species Composition Analysis
An approximation of species composition can be made from the zigzag transect. For example: four shortleaf pines were sampled out of 20 trees, indicating 20 percent shortleaf pine and 80 percent loblolly pine as stand composition.

(vi) Stand Condition Analysis
An approximation can be made of the percent of trees in poor condition in the same manner as used to get species composition. The percentage is not as important as making the landowner aware of the condition of the growing stock. The trees in poor condition can be slated for early removal to favor those in better condition.

Transect information can reveal treatment needs and alternatives. Transects need not be taken in every field or at every change in forest condition. Each planning job is different. An experienced planner will need to take fewer transects than a less experienced planner.

(b) Strip Sampling

In strip sampling, the sample units are continuous strips of uniform width, spaced at a predetermined distance apart. The width of the strips and the distance between the centerline of the strips determines the percentage of the area sampled. See Figure 636-6 for an example of a typical 10 percent strip sampling layout.

Figure 636-6   **Typical 10 Percent Strip Sampling Layout**

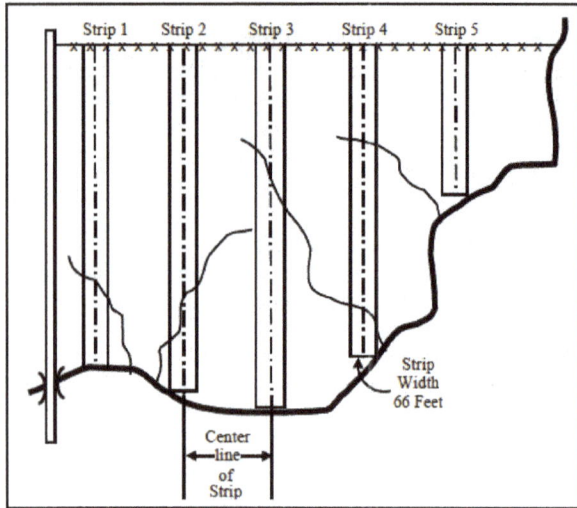

(c) Fixed Plot Sampling

In fixed plot sampling, a set of plots, generally all of the same size, is located throughout the area. The sample plots can be located throughout the area in a number of ways. The most common method is to locate the plots at predetermined intervals on lines a set distance apart. Plots can be any shapes but circular plots are most commonly used. The size and number of plots determines the percentage of the area sampled. See Figure 636-7 for an example of a typical 10 percent line-plot layout.

(d) Variable Plot Sampling

Many features of variable plot sampling (sometimes referred to as "point sampling") and fixed plot sampling are similar. The number and location of plot centers are determined in the same manner (see Figure 636-7). Tree measurements (diameter, height, defects, etc.) are also measured or estimated by methods similar to those used in the fixed plot sampling method.

The primary difference in the two sampling methods is that variable plot sampling does not require measurement of the plot radius because each tree has its own plot size dependent on the diameter of the tree. At each plot center, or "sample point," a count is made of the number of "in" trees whose diameter is large enough to subtend the fixed angle of the angle gauge or prism. See Part 636.3 for details on the use of angle gauges and

prisms. These "sample points" are analogous to the plot centers used in the fixed plot sampling method.

With variable plot sampling, no tree measurements are required if only basal area is desired. If the number of trees or the volume per unit of area is desired, then the dbh or the height of the "in" trees must be tallied.

Figure 636-7   **Typical 10 Percent Line-Plot or Variable Plot Sampling Layout**

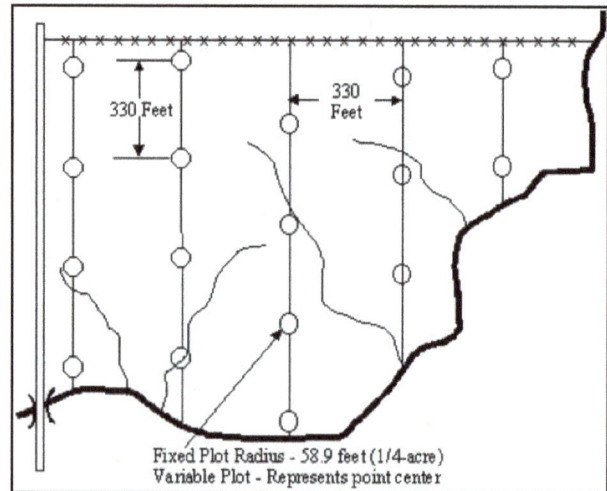

Fixed Plot Radius - 58.9 feet (1/4-acre)
Variable Plot - Represents point center

636.22 Riparian Inventory and Assessment Methods

The NRCS National Water and Climate Center Technical Note 99-1, "Stream Visual Assessment Protocol", is an initial streamside assessment that can be done to provide useful information. It can be downloaded from their web site: www.wcc.nrcs.usda.gov.

Other riparian inventory and assessment information will included as it becomes available.

636.23 Understory Inventory

When an inventory of the production and composition of the understory plant community is needed to adequately plan livestock or wildlife management practices, use the procedures and methods detailed in Chapter 4 of the National Range and Pasture Handbook.

636.24 Windbreak Management Inventory

Reserved for inclusion at a future date.

636.25 Soil Rating Methods

The National Forestry Manual, Part 537.22 lists the rating criteria for several of the common forestry-related and agroforestry-related activities. Each identified soil component in the planning area should be rated to determine its limitation or potential for specific activities. Rating soils for the activities listed in the National Forestry Manual is easily accomplished by using the rating functionality built into the National Soil Information System (NASIS) software. Refer to Part 637.23 for instructions on using this NASIS functionality.

636.30 General

This part provides a detailed description of some of the common inventory tools used in forestry and agroforestry applications.

636.31 Age and Growth Measuring Tools

(a) Increment Borer

An increment borer is a tool used to obtain a cylindrical cross section from a living tree for studying the annual growth rings. (see Figure 636-8).

Figure 636-8   Increment Borer

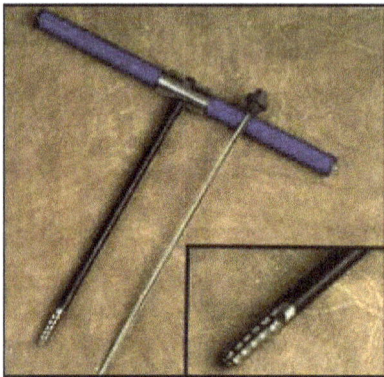

(1) Planning Application
The increment borer is perhaps the best psychological tool for forestry planning and application. Its primary purpose is to determine the age and growth rates of trees. In the planning process, this can be an important function. For example, if a good average diameter for ponderosa pine is 10 inches at 30 years of age, then an examination of the increment core should reveal an average of six rings per inch. If the client has 10-inch trees with more than six rings per inch, it is likely that the trees have too much competition and a thinning or other remedial action may be needed.

(2) Use
The increment borer has a hollow bit (auger) that fits into the handle when not in use. To use, the bit is bored into the tree. An extractor is slipped into the hollow bit to hold the wood core tight. The bit is then reversed a turn, which breaks the core at the tip of the bit. The

extractor is then removed to extract the pencil-like core. The core represents a small cross-ring strip from the bark to the pith of the tree from which the annual rings can be examined and counted. (see Figure 636-9). NOTE: You should NOT plug the hole left by an increment borer. Trees are very capable of repairing the small wound by filling the hole with resin to prevent contamination. You should never insert a stick or anything else as this may actually introduce diseases.

Most trees are elliptical in cross section with the pith along the long axis, offset away from the lean of the tree and toward the prevailing winds. The pith can be reached more accurately by drilling along the long axis rather than along the short axis of the tree bole cross section.

Often the slowest growth and the most tightly packed rings are those closest to the bark. To ensure that these are not lost, the borer must be started slowly while being held very firmly. After the bit is into the tree an inch or so it will continue smoothly.

(3) Age Determination
Conifers and porous hardwood annual rings are easily counted. The annual rings of poplar, cottonwood, sweetgum, maple, and other diffused-porous trees are more difficult to discern and may require the use of a coloring agent such as phloroglucinol in order to see the annual rings. Phloroglucinol is available from forestry suppliers. If available, a microscope (using top illumination only) is useful in counting growth rings.

To accurately determine age, bore the tree at breast height (4.5 feet), count the rings, and, if needed, add the number of years estimated for the tree to reach the height at which the boring was made. Age correction factors are listed in the National Register of Site Index Curves, National Forestry Manual, Part 537.

(4) Care and Maintenance
The increment borer is a fine tool and rather expensive. The tip of the bit must be protected from damage. When returning the bit to its case (the handle), guide it between the thumb and index finger so that it will not hit the steel edge.

Keep the tip of the borer sharp and free of nicks. Special sharpening stones are available from forestry suppliers to sharpen the borer bit. Keep the bit clean and oiled with a light oil.

When working with hardwoods, staining and corrosion can be prevented by using beeswax on the bit, which allows the auger to bore more easily.

The extractor should have a leather loop tied to it so that it can be hung around the neck or to a limb to prevent it for becoming damaged.

Figure 636-9   Increment Tree Core

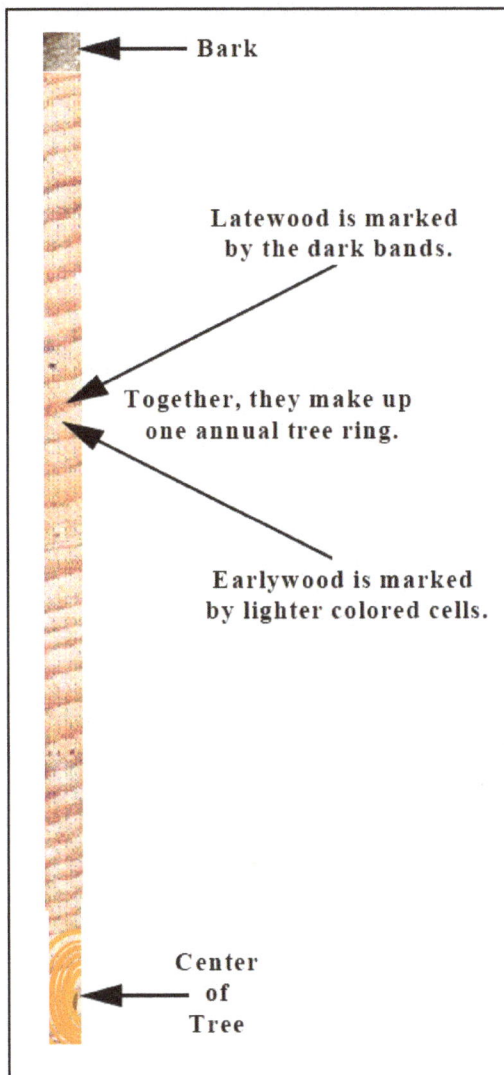

Bark

Latewood is marked by the dark bands.

Together, they make up one annual tree ring.

Earlywood is marked by lighter colored cells.

Center of Tree

636.32 Diameter Measuring Tools

(a) Diameter Tape

A diameter tape is the most popular tool for measuring tree diameter due to its convenient size and relative accuracy. (see Figure 636-10).

A diameter tape is graduated to read 1 inch for each 3.14159 inches of tape length. The principle is that the diameter of a tree is equal to the circumference divided by pi.

Diameters measured with a diameter tape are accurate when properly taken. Care must be taken to hold the tape at a fixed height in going around the tree and to avoid going over bumps or limb stubs.

Figure 636-10   Diameter Tape

(b) Tree Calipers

A tree caliper consists of a graduated bar with one fixed arm set at right angles to the bar, and a second sliding arm that can be adjusted by a screw. (see Figure 636-11).

When placed against a tree, the sliding arm forms a right angle with the bar and is parallel to the fixed arm.

Calipers are accurate for measuring the diameter of standing trees, but are normally limited to trees 36 inches or less in diameter, and are not as convenient to carry as a diameter tape.

636-23

Figure 636-11   Tree Calipers

(c) Diameter Sticks

Several sticks are available for measuring the diameter of a standing tree. These sticks are normally made of wood and contain a diameter (*Biltmore*) scale graduated in such a manner that the diameter of a standing tree may be read directly from the stick. (see Figure 636-12).

To use the diameter stick:

- Hold the stick horizontally against the tree (usually at breast height) and 25 inches from the eye.

- Adjust the stick so that the left end is even with your line of sight to the left of the tree.

- Turn your eye, not your head, and read from the diameter scale the figure crossed by your line of sight to the right side of the tree. This is the diameter of the tree.

Most trees are elliptical in cross section; therefore, it is often necessary to make more than one measurement and use the average to determine the tree diameter.

Figure 636-12   Diameter Stick Measurement

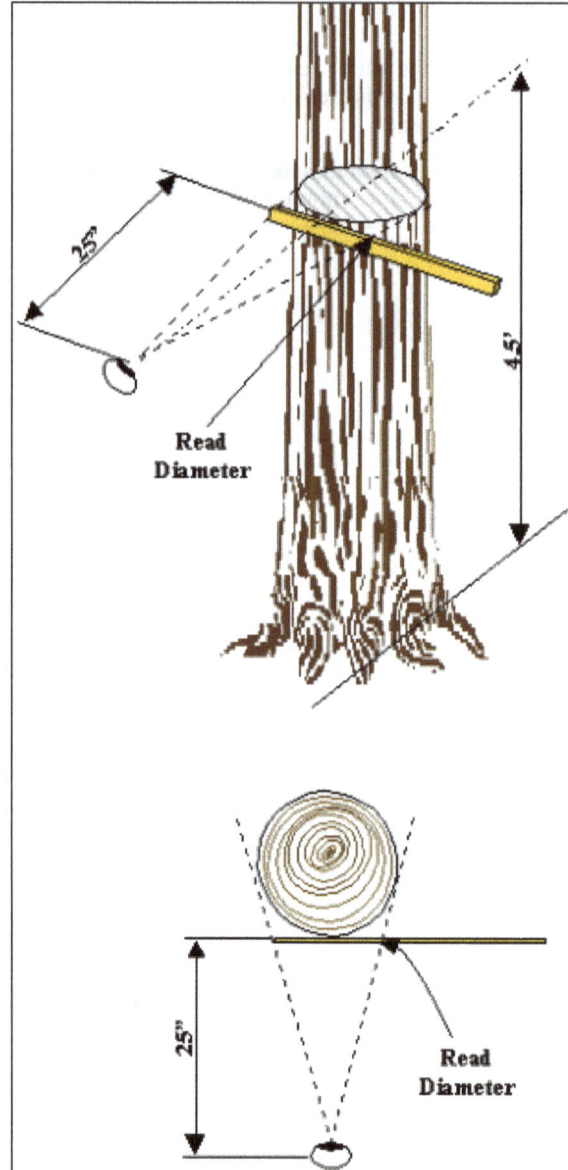

(d) Relaskop

The relaskop combines five instruments into one: a rangefinder for measuring horizontal base distances; a clinometer for measuring tree height; a dendrometer for measuring tree diameter outside the bark; an angle gauge for measuring basal area per acre; and a slope correction instrument for measuring topographic correction. (see Figure 636-13).

Although a very versatile instrument, it is also quite expensive. Consequently, it is not commonly used for the type of activities associated with conservation planning.

Instructions for using the relaskop to measure tree diameters are included with the instrument.

Figure 636-13   Relaskop

636.33 Height Measuring Tools

The height of trees may be measured directly or indirectly. Although it is possible to make direct tree height measurements on standing trees, it is normally most practical to do so only on felled trees.

The most common method of measuring the height of standing trees is with instruments designed to calculate tree height based on indirect measurements.

Tools designed to calculate tree height based on indirect measurements are called hypsometers. They fall into two basic groups: tools based on geometric principles and tools based on trigonometric principles.

Geometric-based instruments, such as the JAL hypsometer, are the standard instrument in some European countries (e.g. Denmark) and are considered quite reliable. The Dendrometer II is another geometric-

based instrument used for quick but approximate measurement of tree height.

Some tools use trigonometric principles to calculate tree heights. For an example of using trigonometric principles to measure tree height, see Figure 636-15.

Several hypsometers are commonly available. Although these instruments have many different names, they all calculate tree height by use of trigonometric principles. These instruments range from the simple and inexpensive Information Stick to the complex and expensive Spiegel relaskop. The following sections describe some of the more frequently used hypsometers.

(a) SUUNTO® Clinometer

The SUUNTO® clinometer (see Figure 636-14) is an instrument that allows you to measure tree heights. When the angle or slope to the top of the tree and the horizontal distance to the tree are measured, the height of the tree can be calculated.

SUUNTO® clinometers are available in several scales, such as degrees, percent, topographic, metric, and secant. With each scale, a different horizontal distance is used. For example, the horizontal base distance is 66 feet for clinometers with a topographic scale and 100 feet for clinometers with a percent scale. Refer to the manufacturers' instructions for the procedure to be used with each scale.

Figure 636-14   SUUNTO® Clinometer

Figure 636-15   Estimating Tree Height Using Trigonometric Principles

Assumptions:
* The tree is truly vertical (i.e. point A is directly above point B)
* The distance to the tree (OC) is the horizontal distance from the operator to the center of the tree
* Leaning trees are measured at right angles to the plane in which it leans

Distance OC is Variable
(as a general rule, the horizontal distance should be between 1 and about 1.5 times the tree height away from the tree).
* Measure distance OC. If the slope is severe, the horizontal distance OC can be measured by holding a measuring tape at point B and stretching it out horizontally until it is exactly above point O.
* Measure angles AOC and COB.
* Determine the lengths of AC and CB by calculator or reference to trigonometric tables.
* Calculate total height (AB)
    * Operator's eye (point O) is above the base of the tree (see example 1):
      AB = OC x (TAN(AOC) + TAN(COB)).
    * Operator's eye (point O) is below the base of the tree (see example 2):
      AB = OC x (TAN(AOC) - TAN(BOC))

Distance OC is Fixed
* Observer stands at a specific distance (or multiple or fraction of it) from the tree (i.e. the distance OC is fixed at 100 feet). The instrument is calibrated in terms of this specific distance and successive angles of elevation and depression, so that AC and CB can be read directly from the instrument.

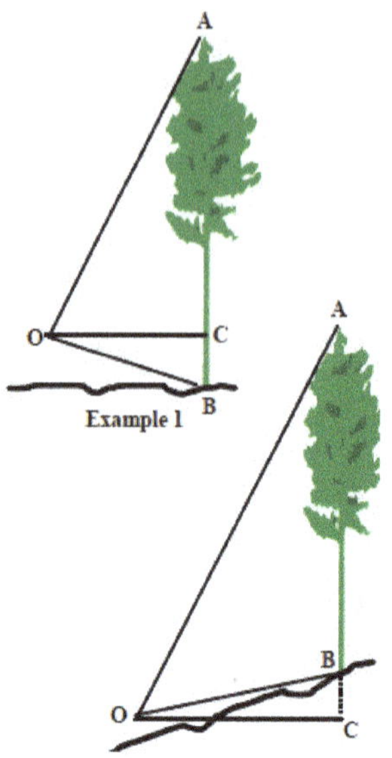

Example 1

Generally, the taller the trees being measured, the greater the horizontal distance should be. The greater the horizontal distance, the easier it is to sight the top of a tree, thus increasing the accuracy of the height measurement. As a general rule, height measurement should be made from a point on the ground such that the angle of observation to the tip of the tree lies between 30 and 45 degrees, i.e. the observer should stand from 1 to 1.5 times the tree height away from the tree.

To use the clinometer, hold it to your eye and with both eyes open, look simultaneously through the lens and alongside the housing. Raise or lower the clinometer (by tilting your head) to place the sighting line at the top or base of the tree.

To measure tree height, stand at the baseline distance, sight the top of the tree and read the scale; sight the base of the tree and read the scale.

To compute tree height, add the 2 scale readings if you looked up to the top of the tree and down to the base, or subtract the height to the base of the tree from the height

to top if you had to look up to both the top and the base of the tree. Multiply this result by the baseline distance to obtain the tree height. (see Figure 636-16).

Figure 636-16   Clinometer Height Measurement

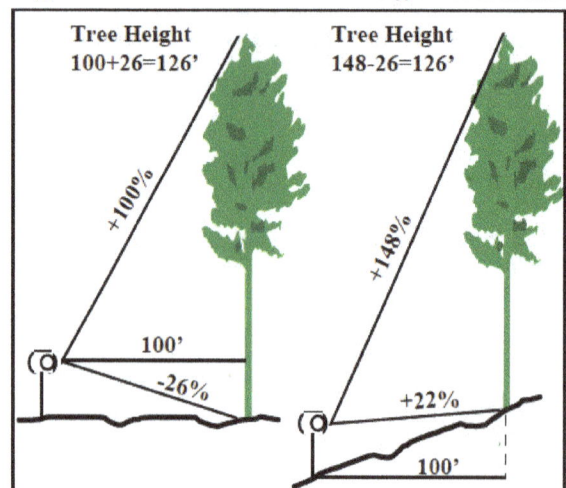

Tree Height
100+26=126'

+100%

-26%

100'

Tree Height
148-26=126'

+148%

+22%

100'

(b) Abney Level

An Abney level (see Figure 636-17) is very similar to a clinometer in the way it is used to measure tree height.

To use the Abney level, sight through the eyepiece to the top of the tree, move the index arm over the scale until the bubble tube is level, then read the scale. Repeat this procedure while sighting at the base of the tree.

To measure tree height, measure the horizontal distance from the base of the tree to a location where the top of the tree can be seen and take readings to the top and base of the tree as explained above.

To compute tree height:

1. If the scale is in percentage, multiply the scale reading to the top of the tree and the scale reading to the base of the tree by the horizontal distance.

Figure 636-17   Abney Level

2. If the scale is in degrees and minutes of the angle, then multiply the horizontal distance by the *Tan* of the angle to the top of the tree and the *Tan* of the angle to the base of the tree.

3. Add the 2 scale readings if you looked up to the top of the tree and down to the base, or subtract the height to the base of the tree from the height to top if you had to look up to both the top and the base of the tree. (see Figure 636-16).

The Haga altimeter (see Figure 636-18) differs from other hypsometers in that it gives the user the choice of selecting several built-in horizontal base distances.

To use the altimeter, look through the pinhole and align the sighting prongs on the top of the tree. Gently press the trigger to lock the pointer. Look through the window and read the scale value indicated by the pointer. Sight on the base of the tree and repeat this procedure.

Figure 636-18   Haga Altimeter

To measure tree height, select a horizontal distance away from the tree corresponding to one of the built-in scale distances, where the top of the tree can be seen. Measure the horizontal distance from the base of the tree to a location where the top of the tree can be seen. Select the appropriate distance scale on the rotating rod and take readings to the top and base of the tree as explained above.

To compute tree height, add the 2 scale readings if you looked up to the top of the tree and down to the base, or subtract the height to the base of the tree from the height to top if you had to look up to both the top and the base of the tree. (see Figure 636-16).

(c) Haga Altimeter

636-27

(d) Relaskop and Vertex Hypsometers

The relaskop is a multi-use instrument that can be used to measure tree height. See Part 636.32 (d) for more information on the relaskop.

The vertex hypsometer (see Figure 636-19) utilizes sonic technology to measure tree height. The unit consists of a hypsometer and transponder. The transponder is fastened to the tree at a predetermined height and the unit calculates the vertical distance from the transponder to the top of the tree. Total tree height is obtained by adding this measurement to the predetermined height of the transponder. Like the relaskop, the vertex is quite expensive and is not commonly used for the type of activities associated with conservation planning.

Figure 636-19   Vertex Hypsometer

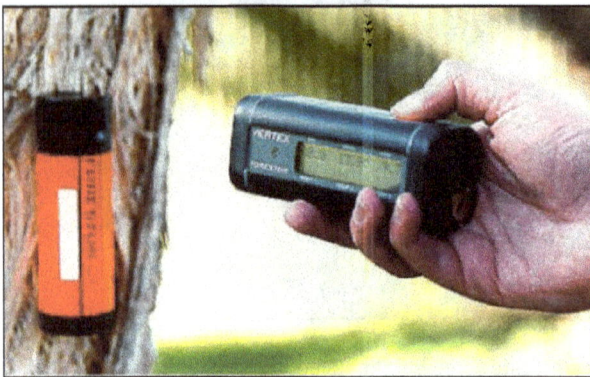

(e) Stick Hypsometers

Several varieties of stick hypsometers are available for measuring tree heights. With each, a graduated stick or rod is held vertically at a certain distance from the eye. Using trigonometric principles (see Figure 636-20), the height of the tree can be determined from reading the scale on the stick where a line intercepts either the top or base of the tree.

The majority of stick hypsometers are constructed to read heights at a set horizontal base distance from the tree (50, 66, 99, and 100 feet) and are designed to align the bottom of the stick with the base of the tree when measuring tree heights.

To measure tree height with this type of hypsometer, hold the stick vertical at 25 inches from the eye and align the bottom of the stick with the base of the tree.

Without moving your head, sight alongside the scale of the stick to the top of the tree and read the tree height. (see Figure 636-20).

Figure 636-20   Standard Stick Hypsometers

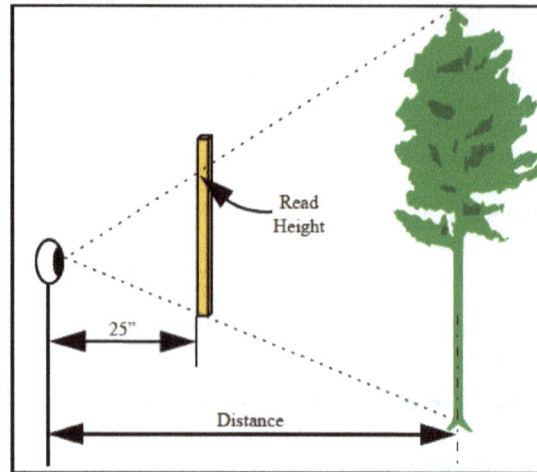

The hypsometer incorporated into the Woodland Information Stick utilizes a graduated percent scale that allows the observer to stand at horizontal base distance rather than a fixed distance. This stick also differs from other stick hypsometers in that it is designed to align the top of the stick with the top of the tree rather than aligning the bottom of the stick with the base of the tree (see Figure 636-21).

Figure 636-21   Woodland Information Stick

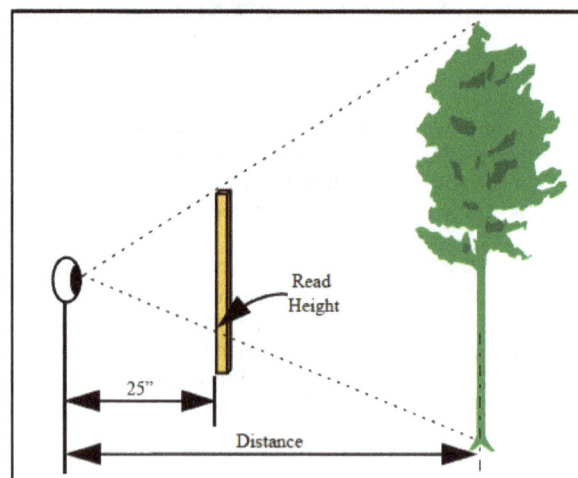

To measure tree height with the Woodland Information Stick, hold the stick vertical at 25 inches from the eye and align the top of the stick with the top of the tree. Without moving your head, sight alongside the scale of the stick to the base of the tree and read the tree height from the inverted scale.

(f) Comparison of Height Measuring Devices

(1) Precision
When properly calibrated and used, the Abney, Haga, Relaskop and Suunto® have equal precision of about 2.5%. The Vertex has a precision of less than 1%. Stick hypsometers are accurate within about 5 to 10 feet.

(2) Ease of Use
The stick hypsometer is the simplest tool to use. The Haga and Vertex are easier to hold steady than the Abney, Relaskop, and Suunto®. The Suunto® causes eyestrain in some people due to the method of projecting the line of sight. Sighting with the Abney and Relaskop is through a tube of narrow dimension or through the instrument. This restricted field of view (especially with the Abney) is important under poor light conditions. The Haga has external sights.

(3) Robustness
Ranked from best to worst:

1. Stick Hypsometer
2. Vertex
3. Relaskop
4. Suunto®
5. Haga
6. Abney

(4) Speed of Use
Ranked from fastest to slowest:

1. Stick Hypsometer
2. Vertex
3. Haga, Relaskop, Suunto®
4. Abney

(5) Price
Ranked from least to most expensive:

1. Stick Hypsometer
2. Suunto®
3. Abney
4. Haga
5. Relaskop, Vertex

(6) Compactness
Ranked from most to least compact:

1. Suunto®
2. Vertex
3. Relaskop
4. Haga, Abney
5. Stick Hypsometer

636.34 Canopy Measurement Tools

(a) Spherical Densiometer

The spherical densiometer is used to determine crown canopy density. (see Figure 636-22).

These data are useful for determining herbage and crown canopy relationships, comparing crown canopy and basal area relationships by species or forest types, and determining stocking percent.

To use the spherical densiometer:

- Hold the densiometer 1 to 1.5 feet away from your body at elbow level. Make sure that the bubble is in the center of its circle, indicating that the densiometer is level.

- Imagine that each of the 24 boxes in the densiometer is subdivided into four quadrants. Count how many of these quadrants contain the reflection of leaves. Record the number.

- Repeat this procedure as many times as necessary to get a representative sample of the area under consideration.

- Average the four measurements. For example:

	Quadrants Covered
Measurement 1	45
Measurement 2	47
Measurement 3	48
Measurement 4	46
Sum	186
Average Quadrants	46.5

- To determine the percentage of canopy cover, multiply the average value by 1.04. For example: 46.5 x 1.04 = 48.4%

Figure 636-22   Spherical Densiometer

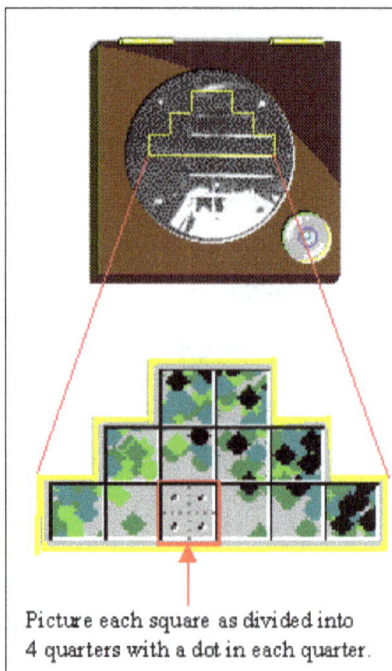

Picture each square as divided into 4 quarters with a dot in each quarter.

(b) Leaf Tube

A simple and inexpensive canopy measuring device can be constructed by attaching two pieces of string across one end of a toilet paper tube to form an X. A third piece of string, about 8 inches long, is tied to the center of the X and extends through the tube, and a weight is attached to the opposite end of the string. (see Figure 636-23).

To use the measuring tube:

- Hold the tube above your head.

- Look up through the tube. Make sure that the weight is directly below the center of the X so you know you are looking straight up.

- If you see leaves touching the center of the X, record a plus sign (+) to indicate that there are leaves directly overhead. If no leaves are touching the X, record a minus sign (–).

- Repeat this procedure as many times as necessary to get a representative sample of the area under consideration.

Figure 636-23   Canopy Cover Measuring Tube

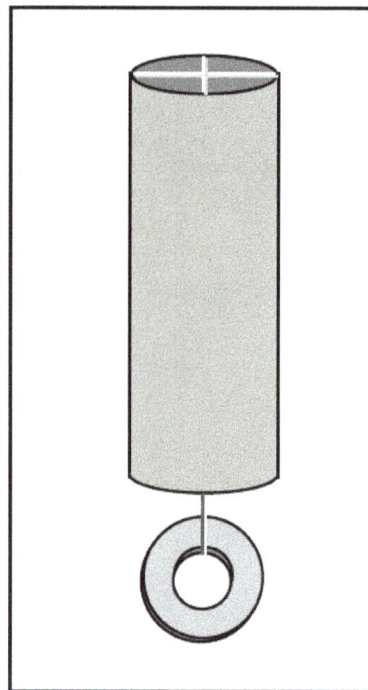

To determine the percentage of canopy cover:

- Sum the total number of plus signs (+) where you observed leaves.

- Sum the total number of pluses and minuses (–) to reflect the number of observations.

- Divide the number of +'s by the total number of observations (sum of +'s and –'s) to determine the fraction of the area which is covered by canopy cover.

- Multiply the fraction by 100 to determine the percentage of the area covered by canopy cover. For example:

  Number of +'s: 18
  Number of +'s and –'s: 49
  Division result: 0.37
  Percentage of canopy cover (x 100): 37%

636.35 Basal Area Measurement Tools

Tools used to measure basal area are useful in determining stand density and are the primary instruments used in variable plot sampling. See Part 636.21(d) for details on variable plot sampling.

(a) Wedge Prism

The wedge prism is a wedge-shaped piece of glass that refracts light rays, thus establishing a critical angle. (see Figure 636-24). A tree viewed through the prism is displaced through an angle depending on the diopter strength of the prism. One prism diopter is equivalent to the right angle displacement of an object by one unit per 100 units of distance.

The diopter of the prism determines the prism's basal area factor (BAF). The most commonly used prisms are those with a BAF between 5 and 40. The larger the average diameter of the stand to be measured, the larger the BAF used. The smaller the average diameter, or the more open the stand, the smaller the BAF used.

Figure 636-24   Wedge Prism

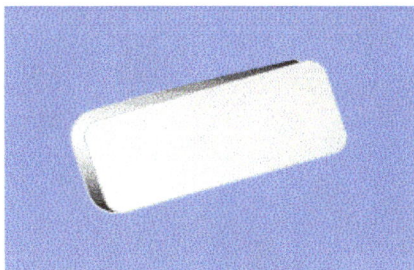

To use the wedge prism, establish a fixed point on the ground. This fixed point becomes the plot center. The prism, and not the cruiser's body, must be held directly over the plot center. The distance the prism is held from the eye is not critical.

While holding the prism at eye level, check each tree in a 360^0 circle, making certain to keep the prism exactly over the center point as you rotate around the plot center. (see Figure 636-25).

Figure 636-25   Fixed-Point Plot Center

As demonstrated in Figure 636-26, when looking through the prism, the edge of the tree is offset to one side. If the offset is greater than the diameter of the tree, the tree is not tallied. If the offset is smaller than the diameter of the tree, the tree is tallied. If the offset is the same as the diameter of the tree, the tree is borderline. Either borderline trees are tallied as half trees, or every other borderline tree is tallied. Another method of determining whether a borderline tree is in or out is to multiply the tree DBH times the plot radius factor (see Table 636 - 1). If the product is greater than the distance from the plot center to the borderline tree, the tree is not tallied.

636-31

Figure 636-26   Wedge Prism Tree Tally

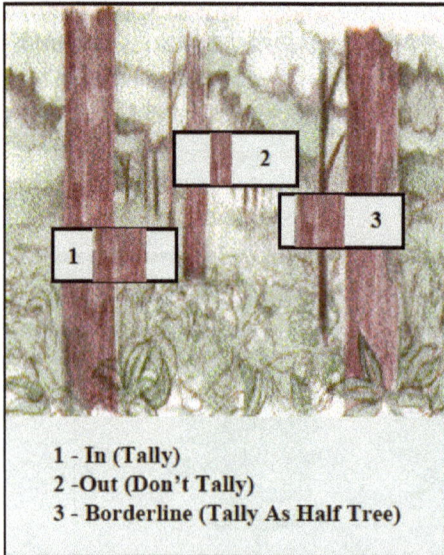

1 - In (Tally)
2 - Out (Don't Tally)
3 - Borderline (Tally As Half Tree)

Table 636 - 1   Plot Radius Factor

Basal Area Factor	Plot Radius Factor
5	3.889
10	2.750
15	2.245
20	1.944
25	1.739
30	1.588
35	1.470
40	1.375
50	1.230
60	1.123
70	1.039
80	0.972

(b) Angle Gauge

There are many varieties of angle gauges. All are based on the principle of projecting an angle to a tree from a fixed point (the eye), through two points whose lateral separation is fixed (the angle gauge). Although there are numerous angle gauges commercially available (see Figure 636-27), a stick-type angle gauge can be simply

constructed of a wooden rod with a peep sight at one end and a crossarm at the other.

To use the stick-type angle gauge, the observer holds the stick against the cheekbone and sights down the rod (see Figure 636-28). All trees whose diameter at breast height (DBH) appears larger than the crossarm are tallied. All trees whose DBH appears smaller than the crossarm are not tallied. All trees whose DBH appears the same width as the crossarm are considered borderline trees. Borderline trees are either tallied as half trees or every other borderline tree is tallied. With a stick-type angle gauge, the observer's eye forms the vertex of the angle; so, unlike the wedge prism, the eye must be maintained over the sampling point as a 360^0 circle is made.

Figure 636-27   Commercial Angle Gauges

A — Fixed length, multiple BAF
B — Fixed length, multiple BAF, slope correction
C — Tube-type, single BAF

Figure 636-28   Stick-Type Angle Gauge

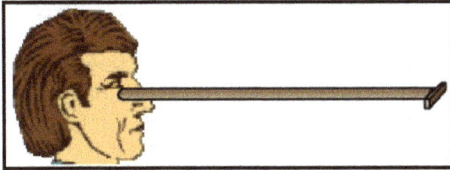

The length of the rod and the width of the crossarm depend on the desired BAF. These measurements are derived by constructing the rod and the crossarm in the same relationship as the ratio of tree diameter to plot radius for the desired BAF, as shown in Table 636 - 2. The ratio of tree diameter to plot radius for a BAF of 10 is 1:33. Thus, an angle gauge with a BAF of 10 could have a 1-inch crossarm at the end of a 33-inch rod.

Table 636 - 2   Tree Diameter to Plot Radius Ratio

Basal Area Factor	Ratio (tree diameter to plot radius
5	1/46.7
10	1/33.0
15	1/26.9
20	1/23.3
25	1/20.9
30	1/19.0
35	1/17.6
40	1/16.5
50	1/14.8
60	1/13.5
70	1/12.5
80	1/11.74

Any combination of crossarm width to rod length could theoretically be used. However, regardless of the ratio desired, the rod should be at least 24 inches long. With rods shorter than 24 inches, it is difficult to keep both the intercept and the tree in focus simultaneously. Thus, if an angle gauge with a BAF of 60 were constructed with a 1-inch crossarm, the rod length would be only 13.5 inches. (see Table 636 - 2).

To adjust the crossarm width to accommodate longer rod lengths, use the formula: $W = KL$ where

W = crossarm width in inches
K = gauge constant
L = rod length in inches

For example, an angle gauge with a BAF of 60 and a rod length of 25 inches would require a crossarm width of:

W = (1/13.5) x 25
W = (0.07407) x 25
W = 1.85

Conversely, to adjust the rod length to accommodate fixed crossarm widths, use the formula: $L = W \div K$

For example, an angle gauge with a BAF of 60 and a crossarm width of 2 inches would require a crossarm length of

L = 2 / (1/13.5)
L = 2 / 0.07407
L = 27

636.36 Linear Measurement Tools

(a) Logger's Tape

Logger's tapes (see Figure 636-29) were originally designed to allow loggers' to scale felled timber. Their design allowed loggers to measure felled timber using one hand by clipping the tape to a belt loop, attaching a hook on the end of the tape to one end of a log and walking to the other end. After the measurement was recorded, a simple yank on the tape detached the hook and the tape automatically rewound.

These features also make this type of tape ideal for many of the inventory methods previously described (measuring plot centers, determining distances between trees, etc.), especially when working alone.

636-33

Figure 636-29   Typical Logger's Tape

(b) Measuring Tapes

Measuring tapes are often used in the inventory process. The primary differences among tapes are (1) material, (2) reel or no reel, (3) length, and (4) unit of measure. Some of the more commonly used tape styles are shown in Figure 636-30.

Figure 636-30   Typical Measuring Tapes

Tape material is either metallic or nonmetallic. Metallic tapes (chrome-clad, epoxy- coated, stainless steel, nylon-coated) are very durable, but the conductive nature of steel can make them dangerous to use around electrical sources, such as electric fences, power lines, etc. Nonmetallic tapes (fiberglass, Dacron) are only slightly less durable than steel tapes and are nonconductive,

although they may be conductive when wet. Therefore, they are generally safe to use around electrical hazards. Although steel tapes are sometimes used without a reel, the convenience of either an open or closed reel is generally preferred. Tapes are graduated in either metric or English units of measurement. A 100-foot tape, graduated in feet and inches, is the style most commonly used.

636.37 Hand Compass

A hand compass is useful in inventory work for locating physical features in the planning area, such as roads, field and forest stand boundaries, windbreak rows, etc.; conducting stand inventories using the plot sampling method; and for general orientation of the planning area to maps or aerial photographs.

There are basically two types of hand compasses, base plate and lensatic.

The base plate compass is a flat, rectangular compass designed to be used in conjunction with a map. Most have a transparent base plate with an etched scale on one or both sides to measure distances on a map. Common scales available are standard map scales (1:24,000, 1:50,000,1:62,000, etc.), inches, and millimeters. Some base plate compasses have a hinged sighting mirror that allows concurrent sighting and viewing of the compass needle, a useful feature when following a course of travel. The base plate compass is the most common type of compass used for general navigation. Figure 636-31 shows examples of base plate compasses, one with a mirror and one without.

Figure 636-31   Base Plate Compass

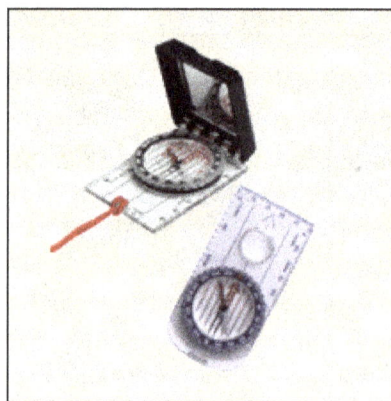

The lensatic compass is generally a round compass with a hinged cover, sighting notch, and sighting wire. One type of lensatic compass, commonly referred to as a pocket transit, functions as a surveyor's compass, prismatic compass, clinometer, hand level, and plumb. The excellent sighting features (especially when used with a staff or tripod) make this type of compass ideal for shooting a bearing, and therefore it is the most common type of compass used for layout work (roads, trails, windbreaks, forest stand boundaries, etc.). Figure 636-32 shows examples of lensatic compasses. The compass on the right is an example of a pocket transit.

Figure 636-32   Lensatic Compass

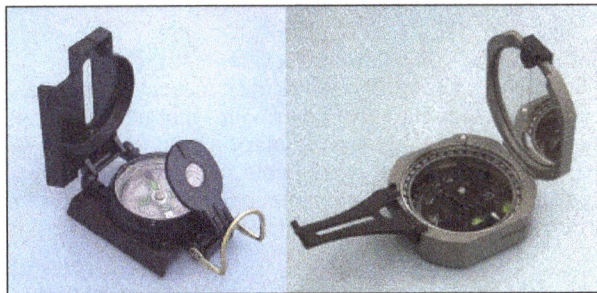

There are other hand compass types available that do not strictly fit into the above two categories. Examples are SUUNTO® and electronic, shown in Figure 636-33.

The U.S. Geological Survey (USGS) has an excellent article on using the compass. The article can be accessed on the Internet at http://mapping.usgs.gov/mac/isb/pubs/factsheets/fs07999.html

Figure 636-33   Other Compass Types

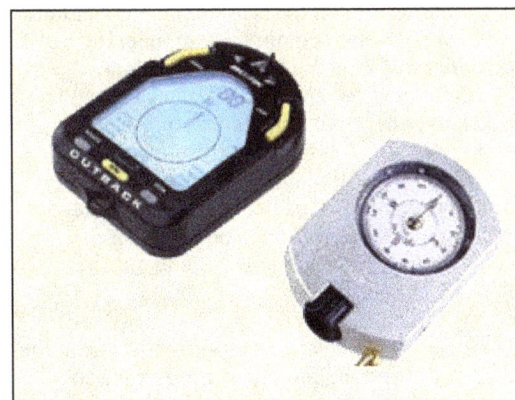

636.38 Wind Meter

Wind meters, also called anemometers, measure wind speeds. These instruments are useful for determining wind parameters used in windbreak designs. Hand-held mechanical and electronic anemometers are the types most frequently used by NRCS personnel. Both are relatively inexpensive and have sufficient accuracy for design purposes. The mechanical type is the least expensive. The electronic types are more expensive but are somewhat more foolproof to use, due to the electronic display. In addition to measuring the current wind speed, some electronic models will also record the average and maximum wind speed over time. Figure 636-34 shows some examples of typical hand-held wind meters.

Figure 636-34   Hand-Held Wind Meters

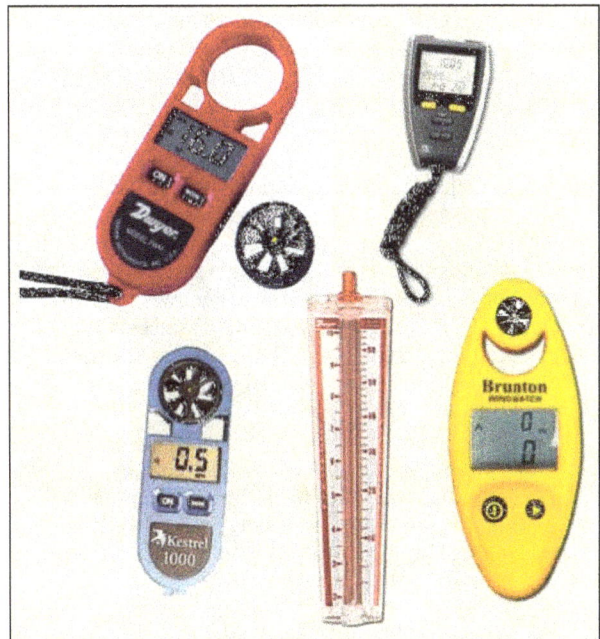

Part 636.4 – Planning Considerations

636.40 General

Success in the planning and application of sound forest management principles depends on an understanding of the ecological processes and the ecology of the communities to be managed. Some processes of change are so universal as to be considered general ecological principles. Others may be less widely applicable (regional) and more closely related to particular communities or individual characteristics of a species.

(a) Dynamics of Ecological Sites

The plant communities for an ecological site are dynamic. They respond to changes in their environment by adjusting the kinds, proportions, and amounts of species in the plant community. Climatic cycles, fire, insects, diseases, and physical disturbances are factors that can cause plant communities to change over time. Some changes, such as those resulting from seasonal drought, are temporary; others, such as indiscriminate logging, may be long lasting.

Other changes are the result of natural successional processes, such as the competition for space between different plant species. This competition for space results in changes to the structure and composition of the forest. As the main story of trees increases in size, the competition for space increases. A few trees gain space because adjacent trees succumb to wind, insects, diseases, logging, or just old age. A few more trees just manage to hold their own. And an increasing majority of trees lose space and eventually succumb.

Thus, many complex factors contribute to changes in the composition, function, and trend of plant communities. Not all changes are related to management practices. Many changes may be caused by climatic fluctuations, fire, and extreme episodic events.

To develop alternatives with the decisionmaker for management of forestland, NRCS employees must understand how an ecological site or association of sites responds to disturbance or other treatment. It is necessary to identify the ecological site and understand the description for that site. The ecological site description has the information necessary to interpret the findings of inventories to determine the rating of an ecological site.

(b) Management Objectives

Management objectives are developed and determined with the landowner during the planning process. All inventory and other necessary information for the development of objectives and the selection and application of forest management practices are gathered during the planning process. The objectives of the landowner and those of the NRCS do not need to be the same, but they must be compatible. The management objective must meet the needs of the landowner and the resources.

(c) Desired Plant Community

For most land units, there are several management alternatives. These alternatives must make available the kind of plant community that provides for and maintains a healthy ecosystem, meets resource quality criteria in the local field office technical guide and the desires of the landowner. The plant community that meets these criteria is the desired plant community.

(d) Identifying Present Vegetation State and Desired Plant Community

An ecological site may have several steady-state plant communities; the actual number is variable from site to site. The common plant communities described in the ecological site description will give an indication of the present plant community and an indication of the factors influencing the transition from one plant community to another.

After the cooperator has set goals for the site based upon the intended use, the NRCS conservationist provides information and analysis to help the cooperator select the appropriate plant community to meet the goals. This plant community becomes the desired plant community (DPC) and the appropriate plans are made by the cooperator to either maintain the existing plant community (if it is the DPC) or plan the appropriate transition from the present plant community to the desired plant community. This decision sets the stage for the selection of the appropriate conservation practices and resource management systems for the cooperator's conservation plan.

The NRCS conservationist will use information from the ecological site description, and other information to assist the land manager. This assistance will provide alternatives that would most likely lead toward the

636-36

desired plant community. This stage of the conservation planning process involves the following steps:

- Inventory the present plant community.
- Determine which vegetation state the present community most closely resembles.
- Determine what changes may be occurring.
- Identify which vegetation state may be the desired plant community to meet the land manager's goals and the resource needs.
- Determine what conservation practice alternatives and resulting resource management system will achieve or maintain the DPC.
- Provide follow-up assistance to the land manager in plan implementation.
- Provide assistance to monitor progress.

Conservation practices applied on forestland are grouped into two categories to reflect their major purposes: vegetation management and facilitating practices.

(e) Recreation

Recreation can be an important use of forestland. Soil interpretations can be used to determine the suitability of an area for recreation. NRCS can assist land users in understanding soil limitations and appraising the physical quality of sites. The major consideration is the protection of the site from erosion.

In analyzing treatment needs, a distinction must be made between areas that are to receive intensive or extensive use:

- On extensively used areas (those walked over occasionally) the best approach is to maintain the natural ground cover, mulch, and litter on the site. The leaves and twigs should not be raked or burned.
- On intensively used areas, such as campgrounds, picnic sites, and playgrounds, a dense groundcover is needed. Materials such as wood chips, asphalt, concrete, and gravel may be used to protect heavily used areas.

Recreational use requires careful consideration of possible conflicts. Certain types of recreation such as camping, nature study, and hiking may require a modification of silvicultural treatments. Trees that are of low priority for timber products may have high priority for recreational areas. A tree that is unmerchantable because of poor form, large limbs, or interior defect may have considerable value as a specimen tree or as a home

for wildlife. An unmerchantable species may have very attractive foliage, bark, or shape. Selection cutting or group selection may be chosen as a harvest system even though clear cutting might be the best method from a wood production standpoint.

Multiple use of forests does not necessarily mean a number of uses on the same area at the same time. Multiple use may involve a series of single uses of the same area over a period of time.

(f) Watershed Protection

Protection of the watershed is an important function of tree cover. A well-managed forest will:

- Reduce flood-producing direct storm runoff.
- Protect the soil from erosion.
- Reduce sediment load in streams.
- Maintain water quality.
- Prolong streamflow during dry periods.
- Permit better water infiltration and percolation to feed permanent springs and streams and replenish the water table.

Little erosion occurs in protected, well-managed forests. Significant erosion may occur along poorly designed haul roads, skid roads, skid trails, and landings even in well-managed forests. Such erosion can be controlled by a few simple but essential measures.

(g) Erosion Control

Proper management can reduce erosion to the geologic rate. One or more of the following measures may be needed:

- Plan roads and trails along the contour to the extent possible. Vary the road grade.
- Slope roadbeds to allow water to drain off the road rather than collect on it.
- Use culverts and bridges and keep equipment out of streams.
- Leave or plant vegetation bufferstrips along streams.
- Use water diversions to dispose of water that collects on roads and skid trails.
- Seed raw roadbanks and other disturbed areas.
- Put road to bed when not in use. Install water diversions, dips, and ditches as necessary to approximate original drainage pattern and seed them with adapted grasses and legumes.

636-37

- Block unused roads to prevent unnecessary traffic.
- Before any harvest operation, plan the work to take into account any moderate or severe soil-related hazards or limitations. For instance, plan to harvest on wet soils during dry season to minimize damage by compaction.
- Regulate grazing in accordance with the technical guide.
- Protect the mat of needles and duff on the forest floor to the extent possible.
- Use geologic information to minimize mass sliding.

(h) Grazable Forests

Some forest communities produce, at least periodically, enough understory vegetation suitable for forage that can be grazed without significantly impairing wood production and other forest values.

Grazing land specialists can assist in preparing guides for grazable forests.

(i) Wildlife

Forests provide habitat for many wildlife species. Silvicultural practices and harvest methods can alter the forest environment. Certain practices are beneficial to certain animal groups and detrimental to others. By knowing animal needs and habitat responses to silvicultural practices, forests can be managed to produce selected timber and wildlife species according to the objectives of the land manager.

Practices that promote diversity in the forests usually produce more variety in wildlife habitat. This can provide a variety of successional stages in the forest area, such as grass-forb, shrub-seedling, pole-sapling, young forest, and mature forest.

(j) Technical Guide

For any planned practice or combination of practices, the technical guide gives the best known specifications. Conservationists must be familiar with and understand the specifications and other important materials in the technical guide. The standards and specifications include practice modifications that can be used to eliminate, overcome, or counteract any moderate or severe soil limitation shown in Section II of the technical guide. If the landowner is being assisted in the application phase by a State-employed forester or a consulting forester, the forester should be given the interpretations and urged to consider appropriate measures.

636.41 Windbreak Planning, Establishment, and Maintenance

Windbreaks may be planted for many purposes or combinations of purposes:

- Act as part of an erosion control system
- Protect crops from wind damage or blowing soil
- Protect livestock
- Protect farmsteads
- Reduce noise
- Provide wildlife habitat
- Give beauty and diversity in areas with few trees
- Control snow drifting
- Keep snow on fields and control deposition
- Conserve energy and feed
- Screen unsightly views
- Enhance recreation areas.

(a) Windbreak Types

Although most windbreaks provide multiple benefits, there are three basic kinds.

(1) Farmstead Windbreak
A row or belt of trees and shrubs established next to a farmstead. Farmstead windbreaks protect soil resources, control snow deposition, prevent physical wind damage to the farmstead, conserve fuel, beautify the area, and provide habitat for some species of wildlife. The chill factor is moderated during winter months, dust is reduced, and living made more pleasant during the summer months.

(2) Feedlot and Animal Shelter Windbreak
One or more rows of trees or shrubs established to protect feedlots or areas where livestock are concentrated. Snow deposition is controlled, thus facilitating feeding and handling. Animal weight gain is better during cold windy weather when so protected. Livestock losses during blizzards are reduced.

(3) Field Windbreak
A row or belt of trees and shrubs planted on field edges or within fields. The primary purposes are to stop soil movement, protect crops, control snow deposition,

636-38

conserve moisture, and provide food and cover for wildlife. Field windbreaks serve best as part of an erosion control system that includes other practice components.

(b) Windbreak Design

Many elements must be considered in designing a windbreak, each of which may be varied to achieve the particular purpose or combination of purposes. Design elements are:

- Species suitability in relation to soils
- Species composition
- Number of rows
- Primary purposes and the desired priority for those purposes
- Species arrangements
- Distance from area to be sheltered
- Spacing of trees within rows and between rows
- Prevailing wind direction
- Topography.

The VegSpec decision support system provides an expert system for the design of windbreaks.

(1) Design for Erosion Control
If wind erosion control is the principal purpose, one or two rows of fairly narrow plantings oriented perpendicular to prevailing winds give good protection.

Windbreaks will give good protection for a distance of 10 times the tree height, fair protection from 10 to 20 times the height, and some protection from 20 to 30 times the height. Windbreaks are most successful in erosion control when used as part of a system that includes crop residue use, tillage practices, wind stripcropping, and cover crops.

(2) Design for Crop Protection
Windbreaks are effective for the protection of orchard and field crops. In fruit crops, windbreaks improve pollination and fruit quality. With crops such as small grains and vegetables, windbreaks minimize the damage from blowing soil.

(3) Design for Livestock Protection
Snow control is most important in feedlots. Placement of the windbreak with relation to feed bunkers must be carefully planned. Other elements are similar to the design for farmstead protection. In planning for livestock protection, do not overlook the possibility of

plantings to keep the dust and odors of the feedlot from reaching the farmstead.

(4) Design for Farmstead Protection
Farmstead windbreaks are most successful when several rows are used. Tall-growing trees are combined with medium-height trees and dense shrubs for maximum protection. Where snow control is important, dense shrub rows are essential on the windward side. Spacing between rows may be varied according to species needs or to fit the cultivating equipment that is available. Conifers generally give better year-round protection. Success with conifers demands special care in planning, spacing, and species arrangements to ensure unrestricted development.

(5) Design for Noise Reduction
For noise reduction, conifers are superior to hardwoods because of their year-round foliage. Better noise reduction depends on maximum foliage density, extra width, and use of tall-growing species. For maximum noise reduction the planting should be placed as close as possible to the source of the noise.

(6) Design for Wildlife Habitat
Windbreaks furnish a necessary habitat element in otherwise treeless areas. Because species diversity is desirable, a mixture of deciduous and evergreen tree and shrub species is best. Where snow is a problem, wider windbreaks are desirable for wildlife survival.

(7) Design for Beauty
To enhance scenic quality, consider the following:

- Mixtures of conifers and hardwoods provide a contrast.
- Silver-foliaged plants when planted next to green foliage produce striking contrast.
- Species with striking fall foliage can be planted where they can be seen.
- Flowering shrubs and trees may be used as parts of rows to add seasonal beauty.
- Trees with unusual color or form may be used where extra room is available to allow optimum development.

(8) Design for Snow Control
A narrow windbreak of medium density is needed to distribute and hold snow on fields. Dense windbreaks are needed to keep snow out of roads, farmsteads, and feedlots.

(9) Design for Energy Conservation

Maximum reduction of winter energy use is achieved with dense windbreaks. Energy savings are proportional to wind reduction.

(10) Design for Screening
Almost any adapted species may be used for screening. If thin-foliage trees are used, it is necessary to back them up with dense-foliage trees or shrubs. For winter screening, conifers are best.

(11) Enhancement of Recreation Values
In recreation areas, windbreaks or screen plantings can beautify as well as provide wind protection.

(12) Design for Odor Suppression

Windbreaks or screening plants that can absorb and disperse air flow from odor generating areas.

(c) Establishment and Management

(1) Site Preparation
Site preparation has two principal benefits. The first is to save moisture through mechanical or chemical fallowing. Nonirrigated plantings in low-moisture areas may require fallowing to ensure enough available water to supply the tree seedlings through the first and most critical summer.

The second benefit of site preparation is to eliminate or control competing vegetation. Fallow systems should leave a crop residue on the surface both to protect from wind erosion and to reduce evaporation.

(2) Spacing
Where moisture is plentiful, spacings of 8 to 16 feet between rows may be suitable. In drier areas as much as 20 to 40 feet may be needed. Often, spacing between rows is selected to fit the equipment available. Spacings within the row may vary from as little as 3 feet for shrubs to 20 feet for trees depending on the species, soil, and climate.

(3) Care of Planting Stock
Planting stock should be stored in a manner that will ensure that the plants remain dormant and retain their vigor. It is important to keep the roots moist from the time the plants are lifted until they are planted.

(4) Planting
Trees and shrubs should be planted with the roots positioned as naturally as possible and near the same depth as they grew in the nursery. The soil must be firmly and uniformly packed around the roots. Any plants that fail to survive the first or second season should be replaced the following planting season.

(5) Care after Planting
No single management practice is more important than controlling weeds and grass. Even when moisture is plentiful, better growth and development occurs with proper maintenance. Where cultivation is necessary, shallow cultivation (4 inches or less) is recommended. In dry areas, or on unfavorable soils, cultivation may be necessary throughout the life of the windbreak or until the crowns close and prohibit cultivation. Herbicides are valuable tools for controlling competing vegetation, especially within the tree row. In dry areas applying supplemental water is beneficial to plant establishment.

(6) Protection
Poultry and livestock must be kept out of the windbreak. Grazing cannot be tolerated if a good windbreak is the goal. Animals can seriously deform and kill trees and shrubs. Insect pests must be controlled. Timely use of pesticides can prevent buildup and provide immediate relief. Long-term solutions include integrating direct silvicultural, biological, and chemical control methods with indirect control methods such as diversity of species and the use of resistant varieties. Periodic inspections are essential to a protection system.

(7) Renovation
Windbreak renovation includes actions to restore or create proper spacing, density, structure, and species composition. There are eight major problems that may be overcome by renovation:

- Need for thinning to relieve extreme crowding and loss of vigor in interior rows.
- Need for releasing conifers from severe overtopping and side competition from broadleaf trees.
- Need for planting supplemental rows of conifers in windbreaks composed entirely of broadleaf trees.
- Need for modifying the level of density of foliage (increase density in some cases and reduce it in others).
- Need for reducing the width of very wide belts where removal of rows will not seriously alter the windbreak's effectiveness.
- Need for managing reproduction so as to maintain the desired structure of the shelterbelt.

- Need for replacing rows lost to drought or disease.
- Need for replacing deteriorating broadleaf rows by underplanting with shade-tolerant conifers.

(d) Selecting Better Trees and Shrubs for Windbreaks

(1) Plant Materials Program
The NRCS plant materials program selects better adapted windbreak plants through testing and release of improved plants. Conservationists and foresters should always be on the lookout for trees and shrubs that have proven to be more drought resistant, more resistant to disease, more resistant to breakage, or have other desirable characteristics.

(2) Research Programs
The Forest Service, Science and Education Administration, universities, and agricultural experiment stations have research responsibilities in windbreak development. It is the policy of NRCS and the responsibility of NRCS conservationists and foresters to work closely with research personnel and thus give them the benefit of field observations. Superior tree selection programs can be developed by close interagency cooperation.

(3) Seed Orchards
After superior trees have been identified, seed orchards and seed production areas need to be established so that seed can be produced in commercial quantities and its origin certified. The goal should be trees grown from seed of certified origin with known genetic potential.

(4) Nursery Production
Trees for windbreak planting are grown in Forest Service, State, private, and conservation district nurseries. NRCS can encourage conservation district governing bodies to develop planting stock orders so as to stabilize the demand. To facilitate production, nurseries should be provided with reasonable estimates or commitments two or more months before the planting season. A good windbreak program requires a variety of species. Planting a few species in great quantity leaves the program vulnerable to disease or insect infestation.

(5) Handling of Planting Stock
Many windbreak plantings require only a few hundred trees. For economy of handling it is necessary to pool orders. NRCS encourages conservation districts to make pooling arrangements. When pooled orders arrive,

modern storage facilities are essential if the trees are to be kept dormant until ready for planting. Where bad weather can delay a planting season or where the workload is heavy, refrigerated storage pays off.

(e) Technical Materials

Each conservationist and forester should make certain that the principles in this manual are contained in the technical guide. The National Forestry Manual, Part 537 discusses conservation tree/shrub interpretations. The conservationist needs to be familiar with those interpretations and use them in helping landowners design better windbreaks.

636-41

Exhibit 636-1   Trees Per Acre by Spacing and Diameter

Spacing (D + x) (feet)	Average Tree Diameter (D) (inches)														
	6	7	8	9	10	11	12	13	14	15	16	17	18	19	20
D+15	99	90	82	76	70	64	60	56	52	48	45	43	40	38	36
D+14	109	99	90	82	76	70	64	60	56	52	48	45	43	40	38
D+13	121	109	99	90	82	76	70	64	60	56	52	48	45	43	40
D+12	134	121	109	99	90	82	76	70	64	60	56	52	48	45	43
D+11	151	134	121	109	99	90	82	76	70	64	60	56	52	48	45
D+10	170	151	134	121	109	99	90	82	76	70	64	60	56	52	48
D+9	194	170	151	134	121	109	99	90	82	76	70	64	60	56	52
D+8	222	194	170	151	134	121	109	99	90	82	76	70	64	60	56
D+7	258	222	194	170	151	134	121	109	99	90	82	76	70	64	60
D+6	303	258	222	194	170	151	134	121	109	99	90	82	76	70	64
D+5	360	303	258	222	194	170	151	134	121	109	99	90	82	76	70
D+4	436	360	303	258	222	194	170	151	134	121	109	99	90	82	76
D+3	538	436	360	303	258	222	194	170	151	134	121	109	99	90	82
D+2	681	538	436	360	303	258	222	194	170	151	134	121	109	99	90
D+1	889	681	538	436	360	303	258	222	194	170	151	134	121	109	99
D+0	1210	889	681	538	436	360	303	258	222	194	170	151	134	121	109
D–1	1742	1210	889	681	538	436	360	303	258	222	194	170	151	134	121
D–2	2723	1742	1210	889	681	538	436	360	303	258	222	194	170	151	134
D–3		2723	1742	1210	889	681	538	436	360	303	258	222	194	170	151
D–4			2723	1742	1210	889	681	538	436	360	303	258	222	194	170
D–5				2723	1742	1210	889	681	538	436	360	303	258	222	194

The number of trees per acre is read at the intersection of the column representing the average tree diameter and the row representing the desired D + x.

CONTENTS

Part 637.4 – Exhibits

637.00 Introduction

This part provides detailed technical guidance and instruction (how-to) relative to the collection, analyses, and development of soil-related interpretations for forestry and agroforestry activities. Instructions and procedures are given for (1) conducting ecological site inventories of forest and windbreak sites; (2) using the

S interpretation generator to automate the
ing of forestry and agroforestry interpretative
s to soil components; (3) using the Ecological Site
ory application to record and query forest and
reak site inventory data; (4) using the Ecological
escription application to record and query
and ecological site descriptions.

irpose of this part is to define how NRCS collects
terprets soil-related forestry and agroforestry
ation. This information is needed to make
ons on the use and treatment of soil, vegetation,
ater resources. Knowledge of the relationship of
forestry and agroforestry activities is essential to
le planning and applications assistance.

01 Policy

to the National Forestry Manual.

02 Responsibilities

to the National Forestry Manual.

03 Basic Unit of Interpretation Forms of Information Display

3asic Unit of Interpretation

isic geographic or land unit for interpretation is the
imponent. For the purposes of this manual, the
soil component normally refers to a soil series or phase
of a soil series that exhibits consistent behavior in terms
of growth, management and response of trees and
associated understory. Soil map unit delineations in
published soil surveys usually contain one or more
predominant soil components and a number of
components as inclusions. Regardless of complexity,
interpretations and supporting data have a single soil
component as their base unit or point of focus.

(b) Forms of Information Display

Soil map unit descriptions and ecological site
descriptions are the two major forms of display used to
describe the information and interpretations known
about individual soil components.

Soil map unit descriptions usually show the anticipated
behavior or limitations of each soil component included
in the map unit.

Ecological site descriptions show group-level
interpretations for soil components that behave similarly;
where necessary, they also include component-level
interpretations for individual soil components. For
example, a group of similar soil components may have
the same interpretation for overstory tree species but
have different erosion hazard ratings. A group-level
interpretation, when used and appropriate, allows for
simplification of database relationships and forms of
display.

637.10 General

Certain data must be collected, analyses made, and evaluations performed to accurately describe the behavior and limitations of soil components for the purposes of forestry and agroforestry. Interpretations associated with each soil component are (1) developed from the raw field data and subsequent analyses, (2) inferred from historical data, maps or anecdotal information, or (3) derived from criteria based on soil characteristics, soil-moisture relationships, and other associated attributes.

Certain interpretations are highly dependent on the analyses of field data, e.g., Trees to Manage; Forest Productivity; Forest Understory, Ground Cover and Structure, etc. NRCS foresters are to avoid making such interpretations without adequate and properly collected data.

Other interpretations are inferred from historical data and maps, e.g., Historic Climax Plant Community, or from expert criteria or rating guides, e.g., Conservation Tree/Shrub Suitability Groups. These interpretations are usually not field-data dependent and can be derived from available reference materials or criteria. As such, they are approximations or expectations of an individual soil component's behavior and limitations.

As stated in the National Forestry Manual, Part 537 forestry and agroforestry interpretations are organized into two subparts. The procedures and techniques used

637.20.1 General

Interpretations are developed to predict soil behavior for specific soil uses (timber production) or specific soil management practices (log landings). They help to implement laws, programs, and regulation at local, State, and national levels. They are used to plan broad categories of land use, such as forestry or agroforestry. They are also used to plan specific management practices that are applied to soils, such as mechanical site preparation or haul road construction.

Soil interpretations may be developed at the State, regional, or national level by interdisciplinary teams. Discipline specialists, such as foresters and grazing lands management specialists, are essential to the development of soil interpretations.

to collect, analyze, and interpret data are described relative to the two subparts.

637.11 Organization of Interpretations

(a) Part 637.2 – National Soil Information System (NASIS) Interpretations

This subpart describes the procedures for (1) using NASIS to record vegetation and productivity interpretations; and (2) using the NASIS interpretation generator to develop interpretive ratings based on forestry and agroforestry criteria.

(b) Part 637.3 – Ecological Site Information System (ESIS) Interpretations

This subpart describes the procedures for (1) conducting ecological site inventories of forest and windbreak sites; (2) using the Ecological Site Inventory application to record and query forest and windbreak inventory data; (3) using the Ecological Site Description application to record and query forestland ecological site descriptions.

Refer to the National Soil Survey Handbook (NSSH), Part 617 for detailed guidance and policy relative to the development of soil interpretations.

The National Soil Information System (NASIS) is used to record forestry and agroforestry soil-based interpretations. The interpretative data recorded in NASIS fall within two broad categories: (1) vegetation and productivity interpretations, and (2) criteria-based interpretations.

This subpart describes the procedures for (1) using NASIS to record vegetation and productivity interpretations, and (2) using the NASIS interpretation generator to develop interpretive ratings based on forestry and agroforestry criteria.

This subpart is not intended to provide detailed instructions on the use of the NASIS software, although

637-3

discipline specialists are encouraged to seek training in its use. This subpart is intended to provide discipline specialists with a broad understanding of the NASIS structure and the vegetation and productivity data elements.

637.20.2 Overview of NASIS Structure

NASIS uses a table-oriented approach for the entry and maintenance of data. These tables are grouped into a set of hierarchically related records where each group of tables has a root table to which all other tables in the group are linked.

Figure 637-1 shows this hierarchical relationship. The group of tables named "data mapunit" begins with a root table, named **Data Mapunit.** As shown in Figure 637-1, the root table is always the top table. One or more components in the **Component** table are linked to a single data mapunit in the **Data Mapunit.** As shown Figure 637-1, the component named ADENA in the **Component** table is linked to the data mapunit named TESTCA in the **Data Mapunit** table.

Part 637.21 provides instructions for entering data in the vegetation and productivity tables in NASIS. Part 637.22 provides guidance in the use of the NASIS application to provide criteria-based interpretations.

Figure 637-1   NASIS Hierarchical Table Structure

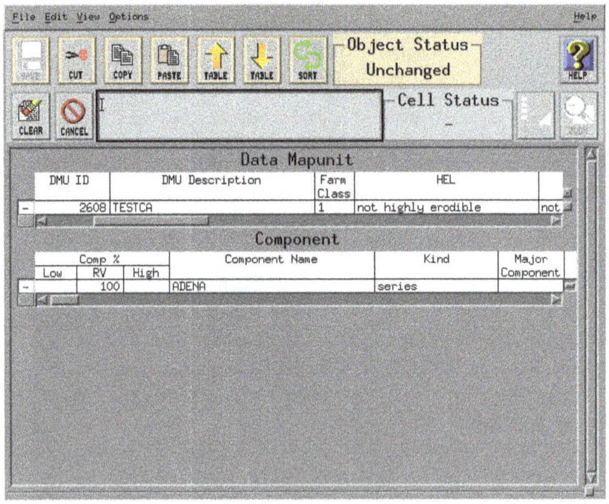

Similarly, several tables are linked to the **Component** table. Figure 637-2 shows the records in one of these tables, **Component Forest Productivity,** that are linked to the component ADENA in the **Component** table.

Figure 637-2   **Component Forest Productivity Table**

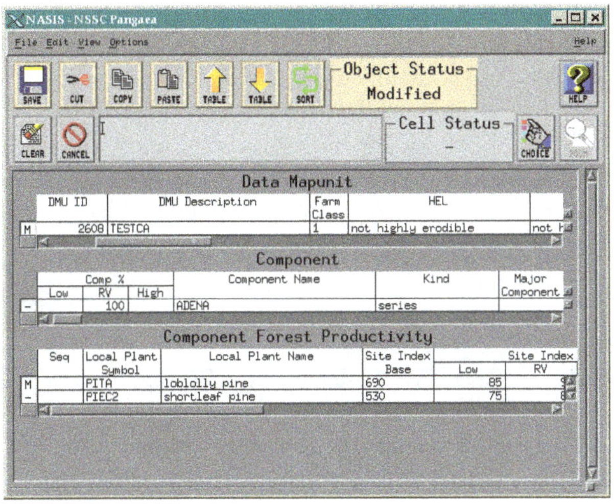

637.21 Vegetation and Productivity Interpretations

All of the vegetation and productivity interpretations are developed from analysis of field data collected for individual soil components. These interpretations correlate plant species and their expected productivity to soil components. Plant discipline experts (foresters, grazing lands management specialists, etc.) are responsible for the collection and analysis of the data. Normally, soil scientists at project, State, or major land resource offices physically enter the data in NASIS, although some areas may choose to allow the plant discipline experts to also enter and/or manage the vegetation and productivity data in NASIS.

All of the vegetation and productivity interpretation records are contained in tables that are linked to the **Component** table. The following sections describe the data recorded in each of these tables.

The National Forestry Manual, Part 537 describes the various vegetation and productivity interpretations and details the data elements associated with each interpretation.

(a) Component Trees to Manage Table

The **Component Trees to Manage** table (see Figure 637-3) contains those tree species that are preferred for planting, seeding, or natural regeneration and/or the

preferred residual tree species in thinning or partial harvest operations.

Figure 637-3   **Component Trees to Manage Table**

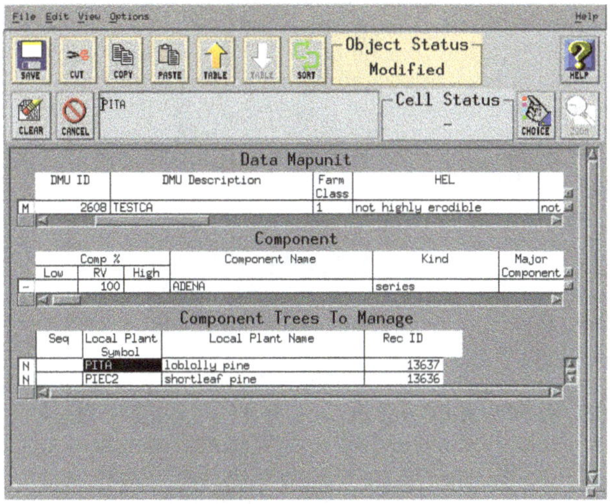

(1) Table Values

(i) *Local Plant Symbol and Local Plant Name*

Select the Local Plant Symbol and Local Plant Name from the list of local plants maintained in the database (see Figure 637-4).

Figure 637-4   **Plant Selection Choice List**

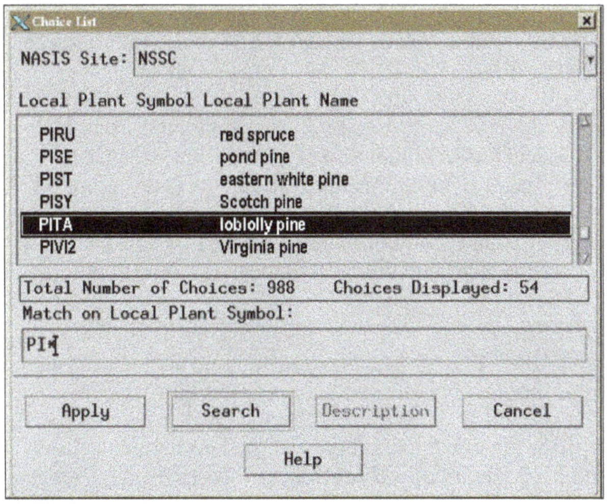

(b) Component Forest Productivity Table

The Component Forest Productivity table contains the site index and annual productivity of adapted tree species. The site index and productivity values recorded (Figure 637-4) are determined from plot data recorded in the Ecological Site Inventory database. See part 637.3 for instructions on the collection of forestry and agroforestry plot data and the use of the Ecological Site Inventory application.

(1) Table Values

(i) Local Plant Symbol and Local Plant Name
Select the Local Plant Symbol and Local Plant Name from the list of local plants maintained in the database (see Figure 637-4).

(ii) Productivity ft3/ac/yr CMAI – Low/RV/High
Productivity ft^3/ac/yr CMAI is the annual growth of identified species in cubic feet per acre at the age of culmination of mean annual increment (CMAI) for the specified site index. These values represent the annual growth for fully stocked natural stands (unmanaged except for protection) as given in standard normal yield tables. An exception to this might be made for species grown primarily in plantations where the use of measured plantation yield seems more logical. Annual growth is the total wood produced in the boles of the trees to the smallest top diameter given in tables. CMAI is the volume at the point of the highest yearly growth. Exhibit 637-1 through Exhibit 637-53 show annual growth in cubic feet at CMAI by site index for common tree species.

Enter the RV (representative value) annual growth in cubic feet per acre that most typifies the site for the identified soil component and site index (see Figure 637-7). Low and high annual growth values (these values are the typical low and high annual growth for the identified soil component and site index) may also be entered.

(iii) Site Index
Enter the RV (representative value) site index, derived from the selected site index base (see Figure 637-6). The site index value entered should be that which most typifies the site for the identified soil component. Low and high site index values may also be entered. These values are the typical low and high site index values for the identified soil component.

(iv) Site Index Base
The site index base identifies the site index curve used to determine the site index. Select the appropriate curve number from the list of site index curves maintained in the database (see Figure 637-5). The list of site index curves is derived from the National Register of Site Index Curves as listed in National Forestry Manual, Exhibit 537-1.

Figure 637-5   **Site Index Base Choice List**

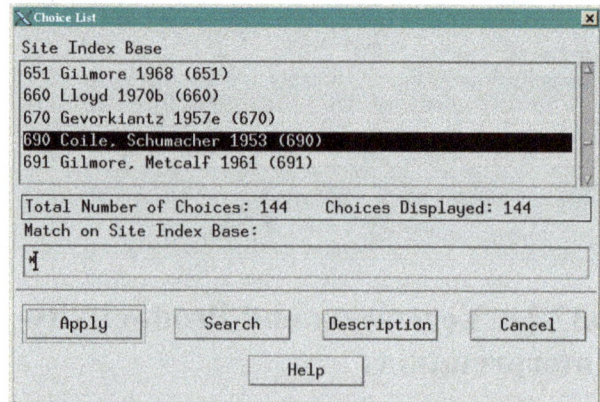

Figure 637-6   **Component Forest Productivity Table - Site Index Values**

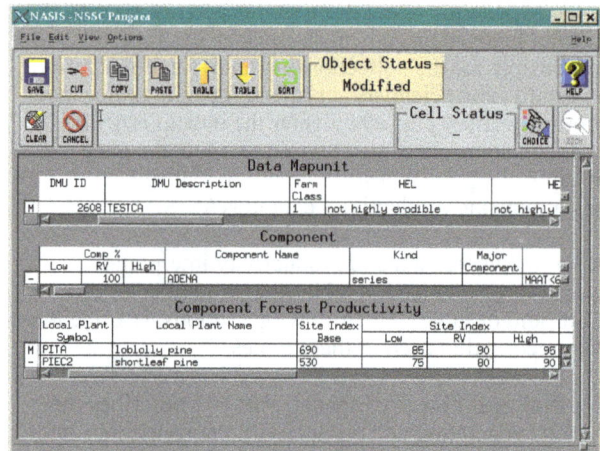

(190-V-NFH, Feb. 2004)

Figure 637-7   Component Forest Productivity
                 Table - Productivity Values

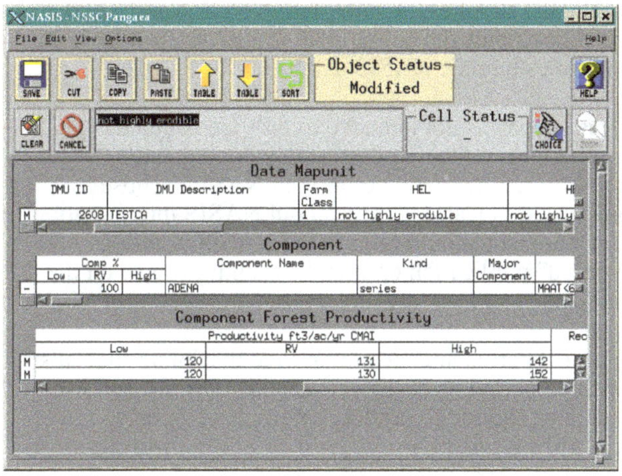

Figure 637-8   Component Forest Productivity
                 Other Table - Site Index

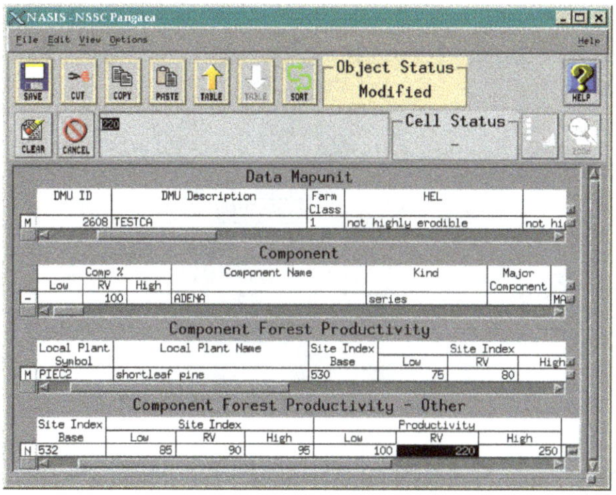

(c) Component Forest Productivity – Other Table

The **Component Forest Productivity – Other** table is used to record site index values with site index bases different from that recorded in the Component Forest Productivity table. This table is also used to record annual productivity values other than cubic feet per acre at CMAI.

(1) Table Values

(i) Site Index Base
Select the appropriate curve number from the list (see Figure 637-5) that identifies a site index curve different than the one used to derive the site index value recorded in the Component Forest Productivity table.

(ii) Site Index
Enter the RV (representative value) site index value derived from the selected site index base (see Figure 637-8). Low and high site index values may also be entered.

(iii) Productivity and Units
Enter the RV (representative value) annual productivity for units of measure other than cubic feet per acre per year CMAI. Low and high annual productivity values may also be entered. Select the appropriate unit of measure from the choice list that corresponds to the entered productivity value (see Figure 637-9 and Figure 637-10).

Figure 637-9   Component Forest Productivity
                 Other Table - Productivity

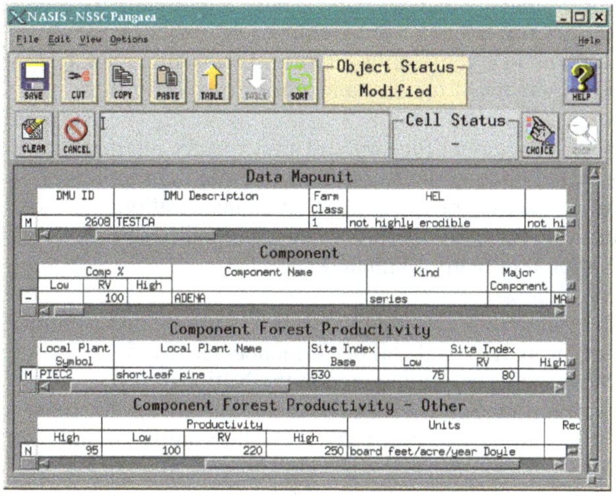

Figure 637-10   Unit of Measure Choice List

Figure 637-10   Unit of Measure Choice List

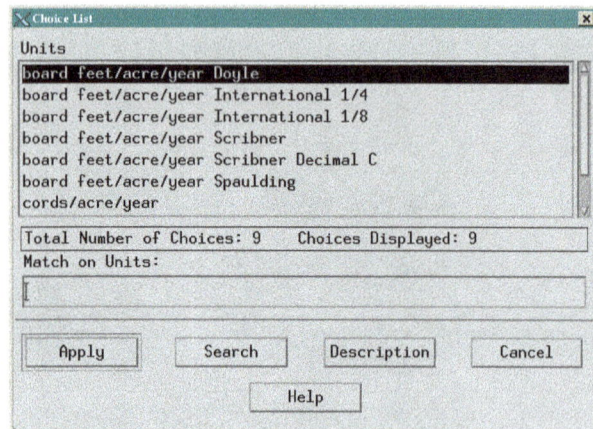

Developing new interpretations or modifying existing interpretations should be done by interdisciplinary teams. Normally, a forester or other plant science specialist supplies the interpretive criteria and a soil scientist inputs the appropriate properties, evaluations, and rules into the NASIS database.

This subpart is not intended to provide detailed instructions on the process of developing and/or modifying interpretive criteria in NASIS. Those interested in learning more about NASIS interpretation development should refer to the NASIS "Getting Started" publication or attend a training course on the subject.

This subpart is intended to provide discipline specialists with an overview of the various components in NASIS that are required to produce interpretive ratings.

637.22 Criteria-Based Interpretations

Criteria-based interpretations are based on soil properties or qualities that directly influence a specified use or management activity. These interpretations predict soil behavior to help in the development of reasonable and effective alternatives for the use and management of forestry and agroforestry activities.

A set of national forestry and agroforestry interpretations is maintained at the national level. The descriptions, rating classes, rating criteria, and considerations for each of these national interpretations are detailed in the National Forestry Manual, Part 537.

Interpretations are assigned to soil map unit components. The properties and soil performance data for these components are maintained in the National Soil Information System (NASIS) database and are the basis for assigning interpretive ratings. Interpretive ratings are assigned to individual soil map unit components by applying the interpretive criteria to the soil map unit component data in NASIS. This process is automated through the standard report functionality in NASIS.

(a) Developing Interpretations

States and other local offices may choose to use some or all of the national interpretations, develop new interpretations, or modify the national interpretations to reflect local criteria and needs. Refer to the National Soil Survey Handbook for guidelines on developing soil interpretations and the policy for documentation.

(b) NASIS Interpretation Components

Three components are required in order to produce automated interpretative ratings using NASIS: properties, evaluations, and rules. All interpretations in NASIS consist of at least one property, one evaluation, and one rule, although most interpretations have several of each. All of the properties, evaluations, and rules used to produce interpretive results are stored in the NASIS database. Anyone with the proper permissions may enter new properties, evaluations, and rules or modify existing ones.

(1) Properties
A property retrieves specified soil data from the database. It consists of a statement written in a language which queries the database and retrieves specific data. Some of the statements are very simply but most are rather complex. Because of their complexity, most properties are developed and/or modified by individuals thoroughly familiar with the language.

Figure 637-11 shows an example of a simple property statement that retrieves the low, high, and representative values for the slope of a soil map unit component.

Figure 637-12 shows the complexity of a typical property statement. This property retrieves the minimum top depth to a soil layer with soil moisture status of "wet" in any month corresponding to a specified taxonomic temperature regime. This property statement demonstrates not only the complexity of the statements but also the level of knowledge of soil

637-8

properties that is required to retrieve soil data relative to specific interpretive criteria.

Figure 637-11   Simple Property Statement

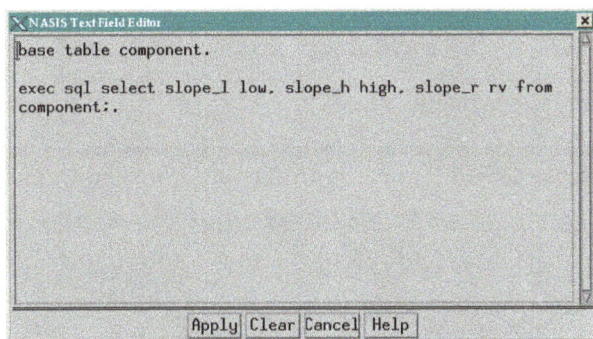

Figure 637-12   Complex Property Statement

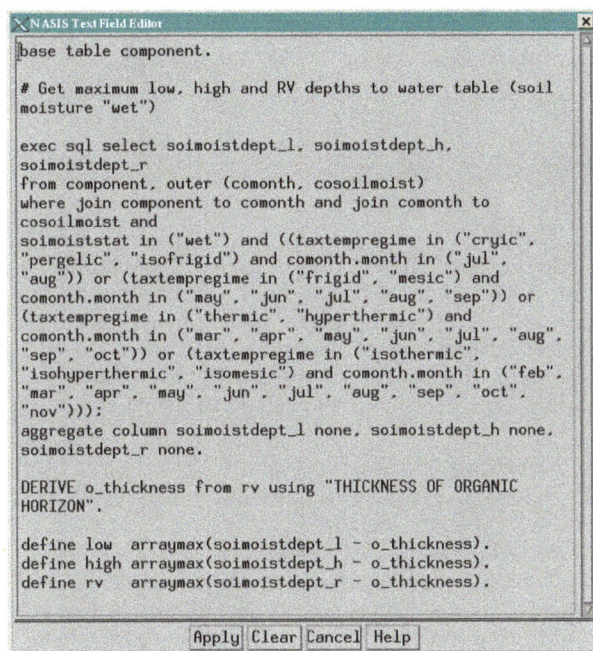

(2) Evaluations

An evaluation compares the data values retrieved by a property to the criteria specified in an interpretation and assigns a numerical rating (rating value) to the result. There are two basic types of evaluations: "crisp" and "fuzzy." The following examples demonstrate the differences between these two evaluation types.

(i) Crisp Evaluations

Crisp evaluations return a rating class value of 1 or 0. In other words, the evaluation result is either true (1) or false (0). For example, an interpretation may rate a soil as being limited for a specific use if the soil is poorly drained. To evaluate the drainage class of a soil, a property is developed to retrieve the drainage class and an evaluation is developed to determine if the drainage class is poorly. If the drainage class is poorly, the evaluation returns a rating value of 1 (true). If the drainage class is not poorly, the evaluation returns a rating value of 0 (false). This type of evaluation is termed crisp. Figure 637-13 shows how the NASIS evaluation editor is used to develop a crisp evaluation.

Figure 637-13   Crisp Evaluation

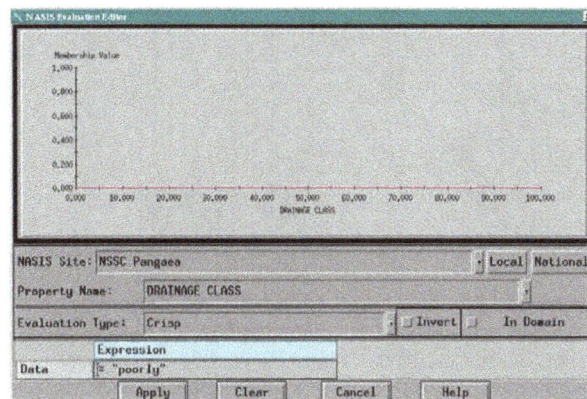

(ii) Fuzzy Evaluations

Fuzzy evaluations return a rating class value of 1 (true) or 0 (false) or some value between 1 and 0. Like a crisp evaluation, a fuzzy evaluation can be either true or false. But, unlike a crisp evaluation, the fuzzy evaluation may also be partly true or partly false. For example, an interpretation may rate a soil as being limited for a specific purpose if the slope is too steep. To evaluate the slope of a soil, a property is developed to retrieve the slope and an evaluation is developed to determine if the slope is too steep. The interpretation defines too steep as any slope over 35 percent. Conversely, the interpretation defines not too steep as any slope less than 5 percent. Therefore, a slope greater than 35 percent has a rating class of 1 (true, or too steep) and a slope less than 5 percent has a rating class of 0 (false, or not too steep). Slopes between 5 percent and 35 percent are neither too steep nor not too steep and have a rating value somewhere between 0 and 1. The steeper the slope, the higher the rating value. The flatter the slope, the lower the rating value.

Figure 637-14 illustrates the fuzzy rating value principle. Note that in the graph, a slope of 20 percent has a rating value of 0.6. A rating value of 0.6 indicates the slope is neither too steep nor not too steep. However, because the value is greater than 0.5, it can be said that the slope is closer to being too steep than it is to being not too steep.

Figure 637-14   Fuzzy Rating Value

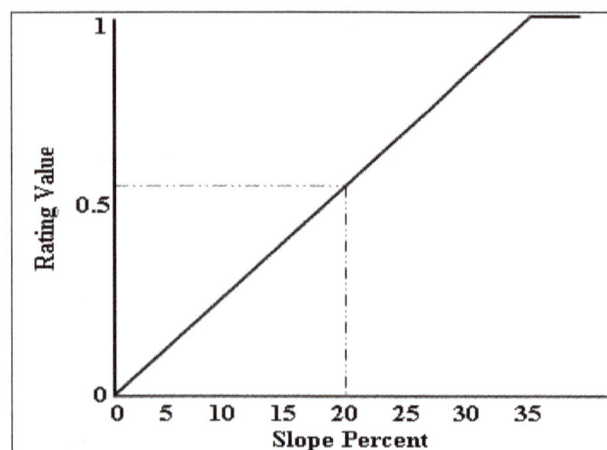

Figure 637-15 shows how the NASIS evaluation editor is used to develop a "fuzzy" evaluation. This example of and evaluation uses a linear curve, but several types of curves are available for use in the evaluation editor. The type of curve used in an evaluation is at the developer's discretion.

Figure 637-15   Fuzzy Evaluation - Linear Curve

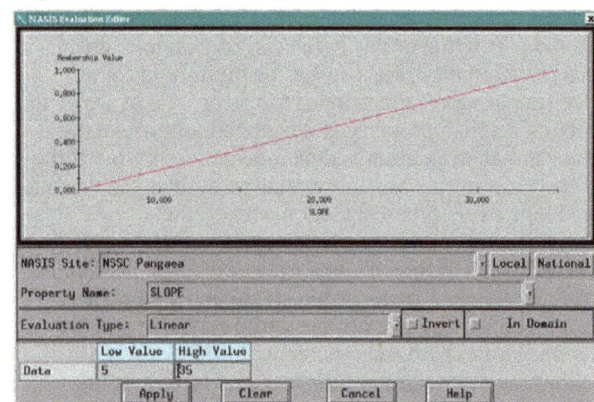

The type of curve used depends, in large part on the developer's determination of how much effect a change in criteria would have on the rating value.

Figure 637-16 demonstrates the use of the sigmoid curve. This curve type gives greater weight to slope changes toward the middle of the slope and lessens the impact of slope changes at the two extremes. Notice in the example, that changes in the slope percent from 5 to 25 percent produce relatively large increases in rating values, while changes in slope percent from 10 to 15 percent and 25 to 35 percent produce relatively small increases in rating values.

Figure 637-16   Fuzzy Evaluation - Sigmoid Curve

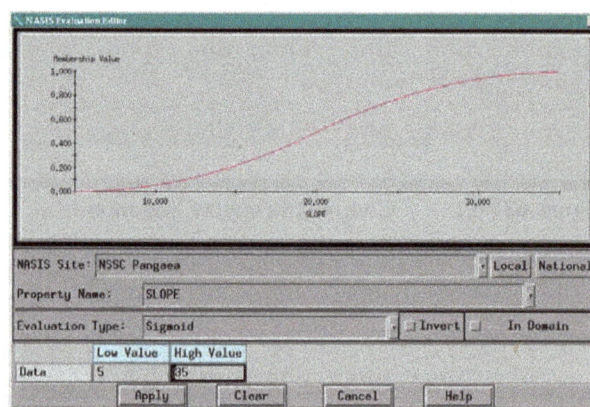

(3) Rules
A rule is used to report out the results of evaluations. Some rules report out the results of only one evaluation, while other rules report out the results of the interaction between several evaluations and/or other rules. Regardless of the complexity of a rule, the end result of a rule is the reporting out of evaluation results.

(i) Reporting Out Rating Values
The following example illustrates how a rule reports out the results of an evaluation. In this example, a rule to report out the results of the evaluation drainage class would look like the one in Figure 637-17. Because the drainage class evaluation is crisp, the rule will report out a rating class of either 0 or 1, meaning the soil is either poorly drained (1) or not poorly drained (0).

Figure 637-17   Simple Rule

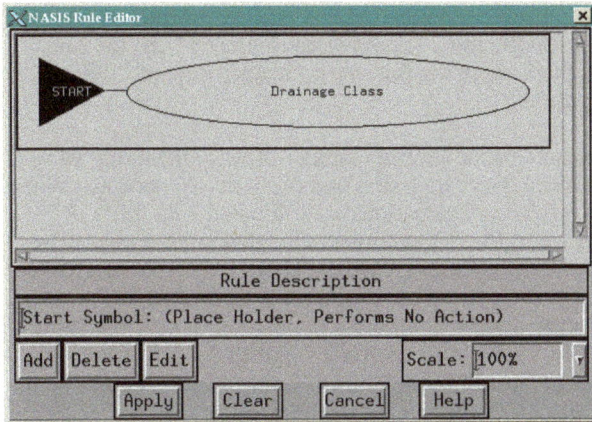

Figure 637-18 shows the result; notice in the report that the rating value is 1.00. A rating value of 1.00 indicates that the result of the evaluation is true and the soil is poorly drained. In this report example, only the rating value (1.000) is reported out.

Figure 637-18   Rule Report - Rating Value Only

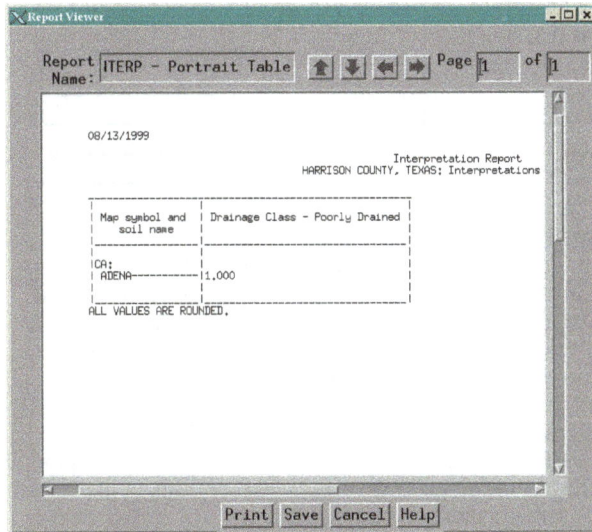

(ii) Reporting Out Rating Classes
In addition to reporting out rating values, most rules also report out a rating class assigned to each rating value. These rating classes are assigned with the NASIS rule editor, as illustrated in Figure 637-19. The assignment of rating classes is determined by those developing the rule. If the rating classes shown in Figure 637-19 are

assigned to the rating values, the resulting report would look like that shown in Figure 637-20.

Figure 637-19   Assigning Rating Classes

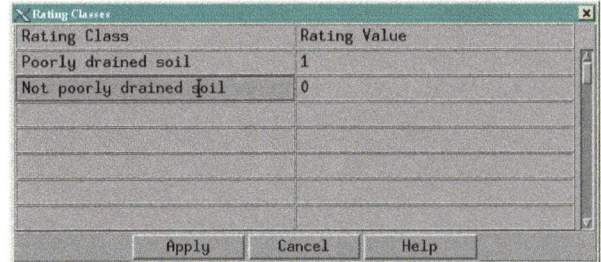

Figure 637-20   Rule Report - Rating Value Class

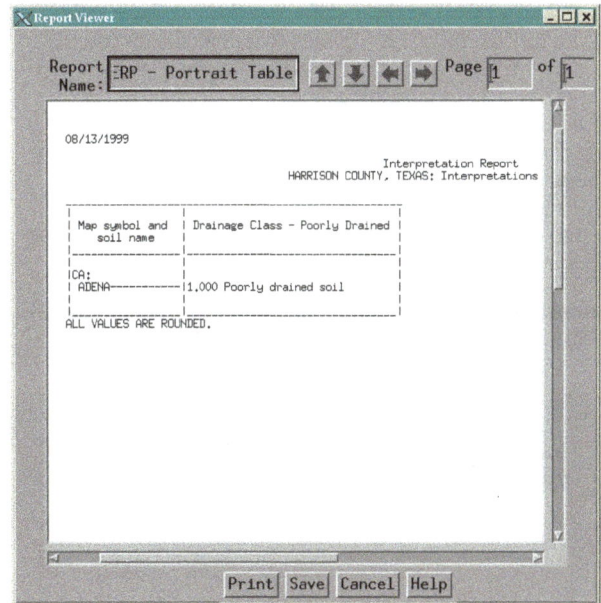

(iii) Rules With Multiple Evaluations
The preceding example illustrates how a rule reports out the results of a single evaluation. Rules also report out the results of the interaction between two or more evaluations. For example, an interpretation may rate a soil as being limited for a specific use if the soil is poorly drained *or* has a slope over 35 percent. A rule to report out the results of the interaction of these two evaluations would look like the one shown in Figure 637-21.

637-11

Figure 637-21   Rule With Multiple Evaluations

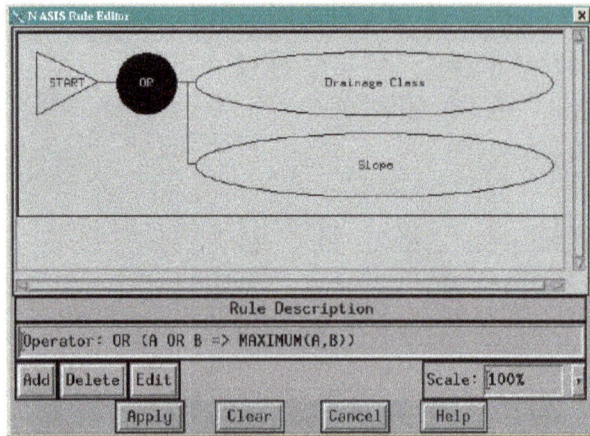

Even though this rule contains two evaluations it will report out only one rating value. The rating value reported out is the result of the interaction of the two rules. This is true of all rules, regardless of the number of evaluations contained in the rule. In this example, if the soil has a drainage class of poorly *or* the slope is greater than 35 percent, the rule will report out a rating value of 1 and the soil will be rated as unsuitable for the specified land use, as shown in Figure 637-22.

Figure 637-22   Rule Report - Multiple Evaluations

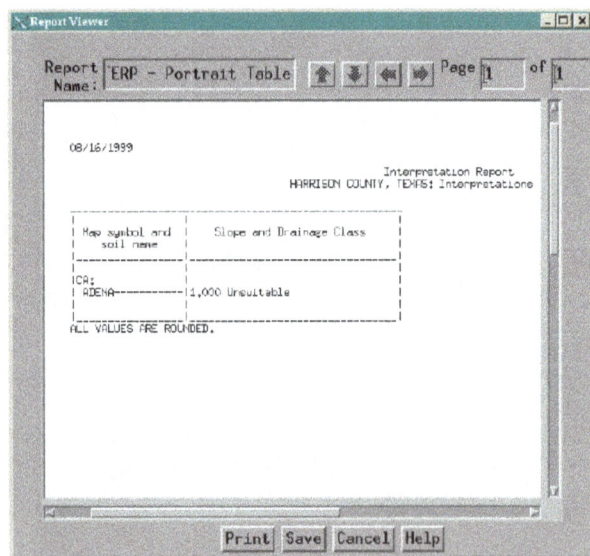

(iv) Fuzzy Math
When a rule contains more than one evaluation, the rating value reported out is a function of the operator used in the rule. As shown in Figure 637-21, the two evaluations in the rule are evaluated using an "OR" operator.

To understand how operators influence the rating value it is necessary to understand fuzzy math concepts. When two or more evaluations are evaluated with the "OR" operator, the rule reports out the maximum rating value of the evaluations (A OR B = Max [A, B]. When two or more evaluations are evaluated with the "AND" operator, the rule reports out the minimum rating value of the evaluations (A AND B = Min [A, B]. Table 637-1 shows the results of the "OR" and "AND" operators using fuzzy math.

Table 637 - 1   Fuzzy Math

If A is:	OR	If B is:	The Rating Value is the Maximum Value:
1		1	1
1		0	1
0		1	1
0		0	0
If A is	**AND**	**If B is:**	**The Rating Value is the Minimum Value:**
1		1	1
1		0	0
0		1	0
0		0	0

(v) Multiple Reporting Levels
Figure 637-22 shows the rule report resulting from the evaluation of the two evaluations, "drainage class" and "slope". A rating value of 1.00 indicates that the result of the evaluation of one or both of the evaluations is true, either the soil is "poorly drained" or the slope is greater than 35 percent, or the soil is both "poorly drained" and the slope is greater than 35 percent. However, there is no way to determine from this report whether the slope or drainage class, or both, is the reason the rating value is 1.00.

In order to report out the individual rating values for rules with multiple evaluations it is necessary to create a rule for each individual evaluation. Figure 637-23 shows an example of a rule constructed to report out rating values for each evaluation in the rule. This rule will produce the exact same results as the rule shown in Figure 637-21.

Figure 637-23   Rule With Multiple Rules

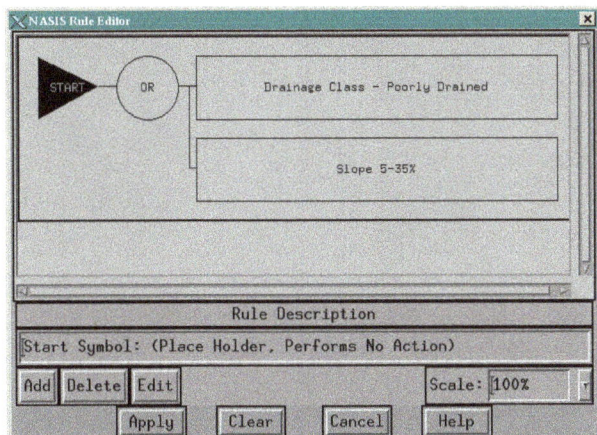

The difference between the rule shown in Figure 637-21 and the rule shown in Figure 637-23, is that the rule shown in Figure 637-23 consists of two "child" rules rather than two evaluations. As shown in Figure 637-24, because this rule consists of "child" rules (Drainage Class - Poorly Drained and Slope 5-35%), rather than evaluations, the rating values assigned to each of the "child" rules can be reported out, in addition to the rating value for the rule as a whole.

Figure 637-24   Rule Report - Multiple Reporting Levels

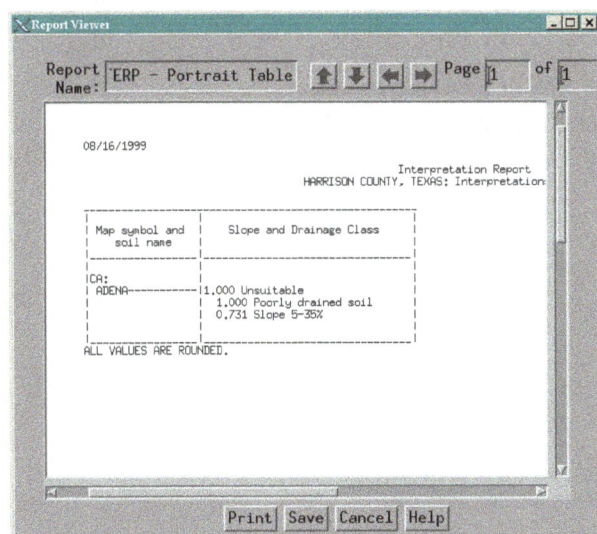

All of the forestry interpretations in NASIS are constructed using "child" rules. This method allows the

user to report out as little, or as much detail, as needed for the particular application. Figure 637-25 shows a typical forestry interpretation report with all "child" rules reported out.

As shown in Figure 637-24, the rating values for each of the "child" rules is reported out, in addition to the rating value for the rule as a whole. The rating value for the rule as a whole has not changed from that reported out in the rule that consisted of two evaluations (Figure 637-22). Now, however, it can be seen that the rating value of 1.00 is due to the component being "poorly drained" (remember, an "OR" operator returns the maximum rating value). It can also be determined that the slope is less than 35% because the rating value is less than 1.00.

Figure 637-25   Typical Forestry Interpretation Report

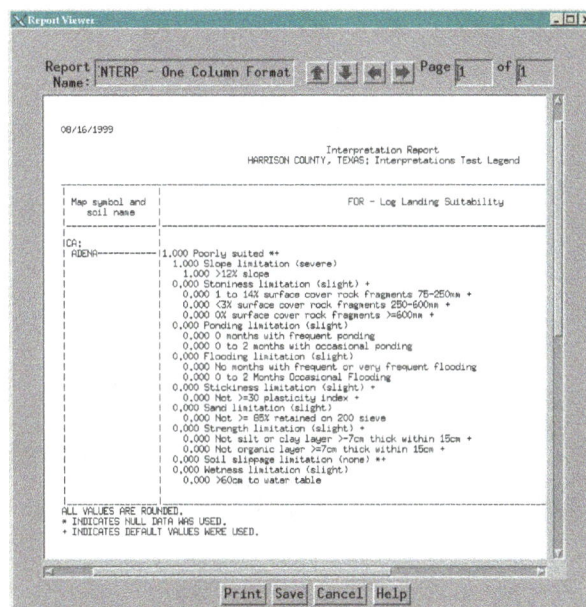

Subpart 637.22(c), below, provides detailed instructions on using the NASIS **Report Manager** to report out interpretations.

(c) Reporting Interpretations

All of the forestry and agroforestry criteria-based interpretations listed in the National Forestry Manual, Part 537 can be automatically generated using the interpretation generation capabilities of the NASIS software. This subpart provides detailed instruction on

637-13

using the NASIS **Report Manager** to report out interpretations.

(1) Report Manager
All the rules stored in NASIS are reported out through the **Report Manager**, including the forestry and agroforestry criteria-based interpretations. For information about rules see Part 637.22(b)(3). To access the **Report Manager**, select **Standard Reports** from the **Options** menu (see Figure 637-26).

Figure 637-26   Accessing the Report Manager

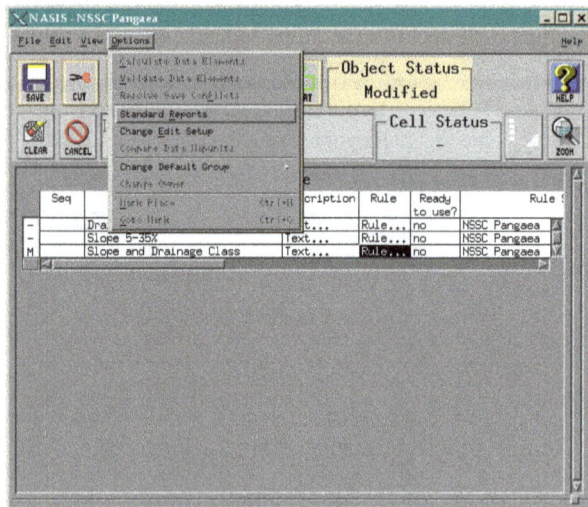

The **Report Manager** screen (see Figure 637-27) is divided into two sections. The section on the left side of the screen contains a list of all the reports available for reporting out. The reports listed depend upon the NASIS Site selected. The NASIS site "NSSC Pangaea" contains the list of national reports that are available to all NASIS users. The NSSC Pangaea site contains all of the forestry and agroforestry criteria-based interpretations, as listed in the National Forestry Manual, Part 537. Other NASIS sites may have developed local reports (including interpretations) that are applicable to their particular locale and supersede the corresponding national report. Consult with the appropriate personnel to ascertain if there is a local report that should be used in lieu of the corresponding national report.

Figure 637-27   Report Manager Screen - INTERP Format

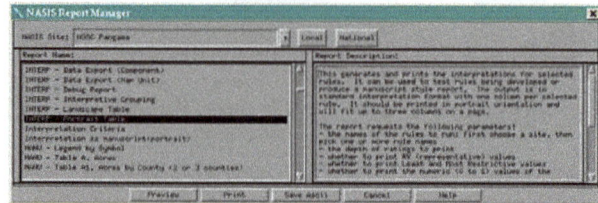

The section on the right side of the **Report Manager** screen contains a narrative that describes the selected report and lists any important information the user should know.

(2) Selecting the Report Format
The report you select in the **Report Manager** determines your report options. There are two principal categories of report formats – INTERP and MANU. Each of these is described below.

(i) INTERP Report Format
The INTERP report format allows you to select any of the rules stored in NASIS. This format gives you the greatest flexibility in determining the report output. You can select a single rule or any combination of rules to report out. With this format you can choose to report out an interpretation (remember, an interpretation is just a rule containing multiple "child" rules) or you can choose to report out only a single "child" rule. Within the INTERP category, three report formats are available:

- INTERP – Landscape Table
- INTERP – Portrait Table
- INTERP – Debug Report

The difference between landscape format and portrait format is the orientation of the report on the printed page. Both of these reports are formatted with one column per selected rule. The portrait format will accommodate a maximum of three columns per page (see Figure 637-28). The landscape format will accommodate a maximum of six columns on a page (see Figure 637-29).

The debug format reports out very detailed information about a rule. It is used primarily by developers to assist in the development and maintenance of rules, but it can also be useful to those needing more detailed information about a particular rule (see Figure 637-30).

Figure 637-28   Portrait Report Format

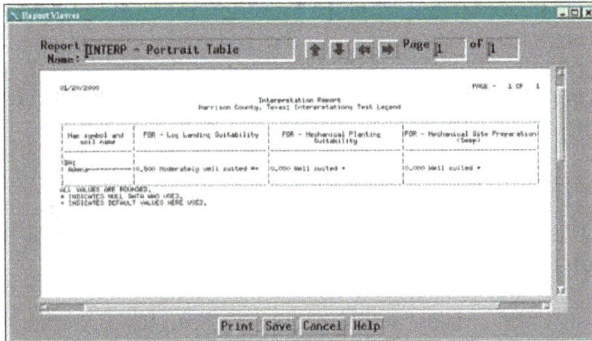

Figure 637-29   Landscape Report Format

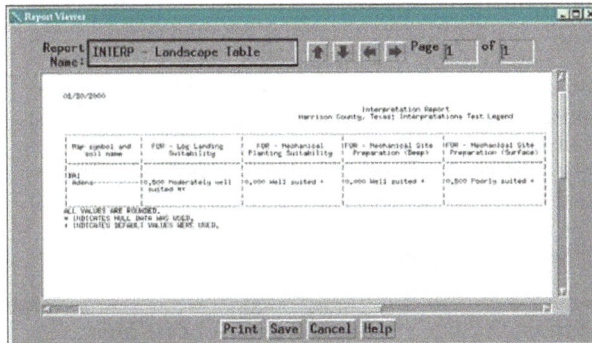

Figure 637-30   Debug Report Format

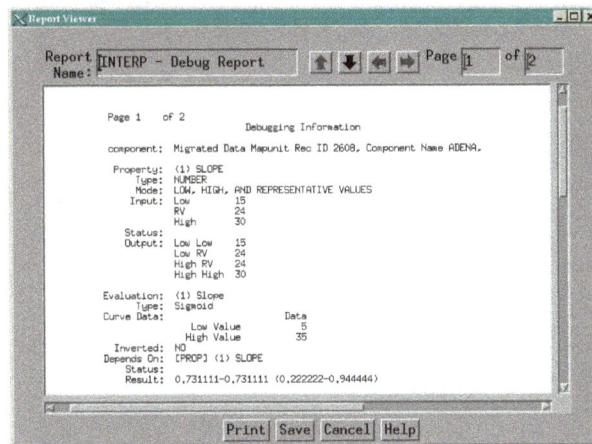

To report out rules using the INTERP category of reports, select one of the report formats listed above from the list in the **Report Manager** (see Figure 637-

27) and click the Preview button. This will bring up the **NASIS Report Parameters** screen. On this screen you select the rule(s) to report out and the various report options (see Figure 637-31).

Figure 637-31   NASIS Report Parameters Screen

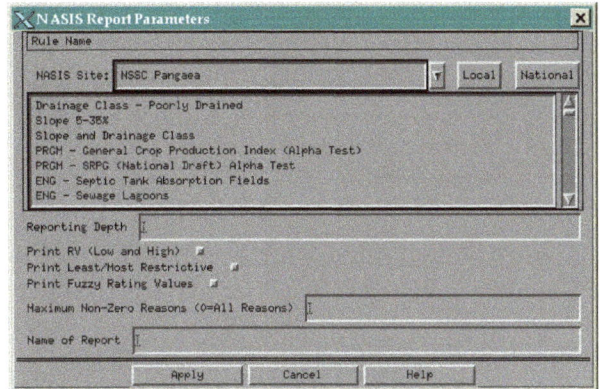

The following options are available on the **NASIS Report Parameters** screen.

NASIS Site – A selection box for choosing the NASIS site from which you want to load existing rules. The default entry is the local site. This site can be selected from the drop-down choice list or by clicking on the Local or National buttons at the right of the selection box.

Rule List – A selection box for choosing the rule(s) to report out. To select a rule, click on the rule name. To de-select a previously selected rule, click again on the rule name.

Reporting Depth – Specifies the number of rules to report out. Some rules contain multiple levels of "child" rules. This option allows you to specify how many levels of rules are reported out. This field must be populated.

Figure 637-32 and Figure 637-33 show how a report for the rule Slope and Drainage Class will look when the reporting depth is set to 1 and 2, respectively. In Figure 637-32, the rating value (1.000) and the rating class (Not Suitable) are reported out only for the "parent rule" (level 1). In Figure 637-33, the rating value and rating class are reported out for both the "parent rule" (level 1) and the two "child" rules (level 2). Notice that each level is indented from the preceding level as an aid in identifying the different levels.

637-15

Figure 637-32   Sample Report - One Level Reporting Depth

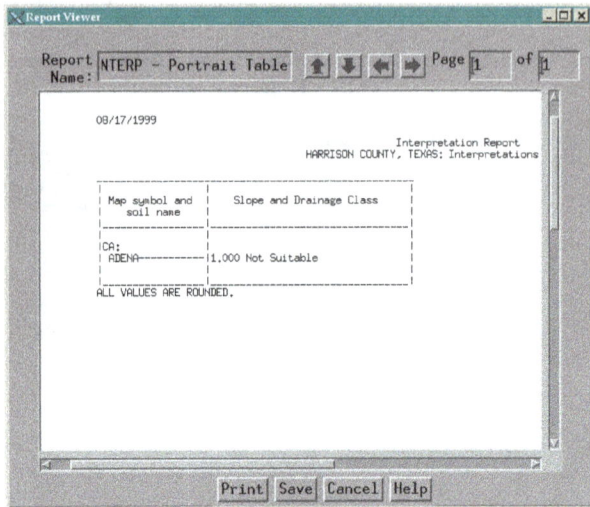

Figure 637-32 and Figure 637-33 show how reports will look if the Print RV (Low and High) is selected and the Print Least/Most Restrictive Feature is not selected. Notice that only one rating value and rating class is reported out. This is because there is only one representative value (RV) for each of the soil properties used in the rule.

Print Least/Most Restrictive – An option to report out both the most and least restrictive rating values. This option allows you to report out the best and worst rating values based on the entire range of soil properties.

If this option is not selected, then the Print RV (Low and High) option should be selected. If neither option is selected, the resulting report will be blank.

Figure 637-34 shows how reports will look if the Print Least/Most Restrictive Feature is selected and the Print RV (Low and High) is not selected. Notice that two rating values and two rating classes are reported out for slope. These represent the range of values for the slope of this soil component in the database. Also notice that only one rating value and one rating class is reported out for drainage class. This is because drainage class is a crisp value and does not have a low and high value (either it's poorly drained, or it's not).

Figure 637-33   Sample Report - Two Level Reporting Depth

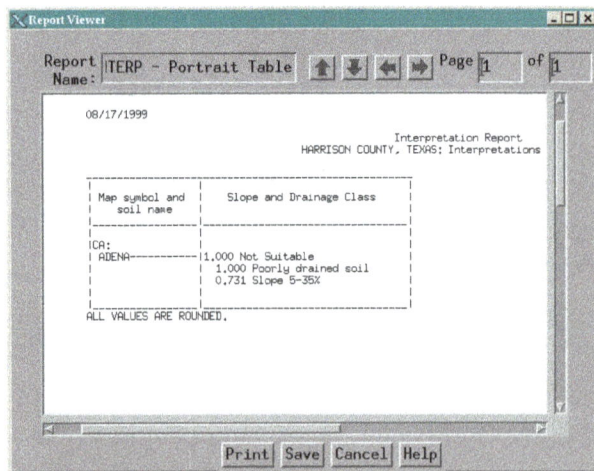

Figure 637-34   Sample Report - Least/Most Restrictive Feature

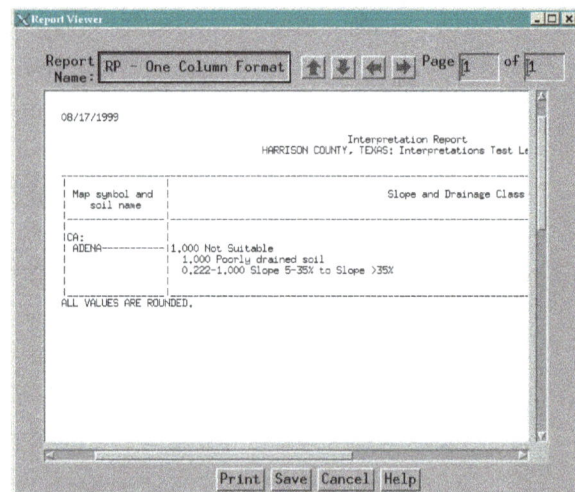

Print RV (Low and High) – An option to report out the low and high representative values when the soil property has multiple RV's, like depth to top of water table. This option allows you to report out the typical range of a soil property.

If this option is not selected, then the Print Least/Most Restrictive Feature option should be selected. If neither option is selected, the resulting report will be blank.

Print Fuzzy Rating Values – An option to report out the rating values between 0 and 1. If this option is not selected, no rating values will be reported out.

If this option is not selected, the "Print Least/Most Restrictive" option should be selected. If neither option is selected, the resulting report will be blank. Figure 637-35 shows how reports will look if the Print Fuzzy Rating Value option is not selected. This is the same report as that shown in Figure 637-33, except that the rating values are not reported out.

Figure 637-35   Sample Report - No Rating Values

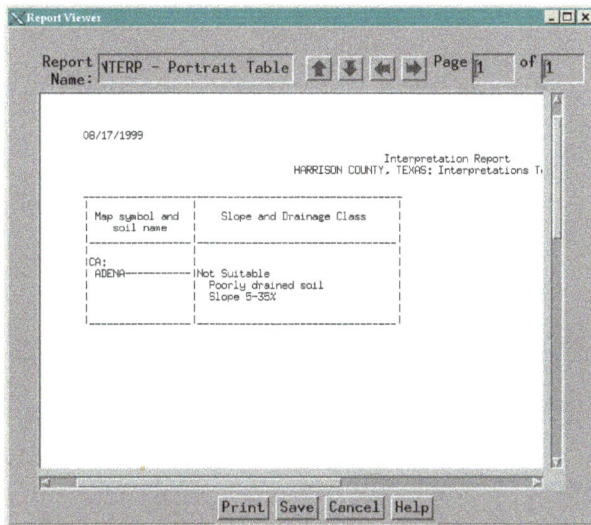

Maximum Non-Zero Reasons (0=All Reasons) – Specifies the number of affecting (non-zero) features reported for each reporting depth.

If you enter a 1 in this field, only the rating value for the "parent" rule will be reported out, even if the rule has one or more "child" rules. Figure 637-32 shows how a rule would look if a 1 is entered in the Maximum Non-Zero Reasons (0=All Reasons) field. Figure 637-36 shows how the same report would look if a 2 is entered in the Maximum Non-Zero Reasons (0=All Reasons) field.

Name of Report – Allows entry of an optional report name.

Figure 637-36   Sample Report – Two Non-Zero Reasons

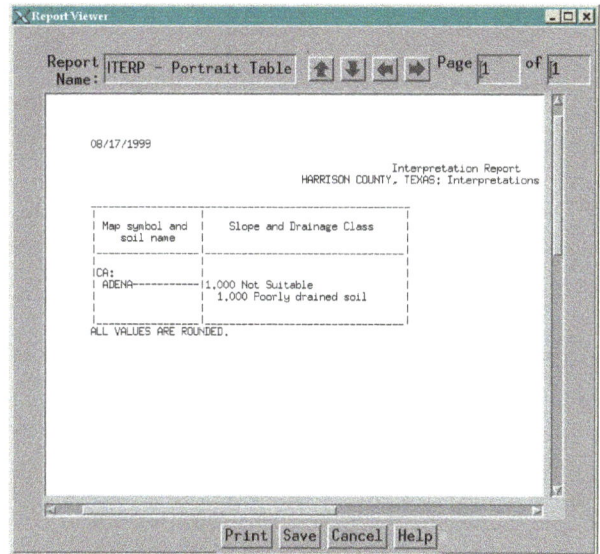

(ii) MANU Report Format

The MANU report format reports out a predetermined set of forestry criteria-based interpretations. The reports have been formatted to produce a manuscript style report for publication. The output is in standard interpretation format with one column per selected rule. The report contains three interpretations on a page and should be printed in portrait orientation. The MANU format reports are selected from the list of reports in the **Report Manager** (see Figure 637-37). An example of a MANU report is shown in Figure 637-38.

Figure 637-37   Report Manager Screen – MANU Format

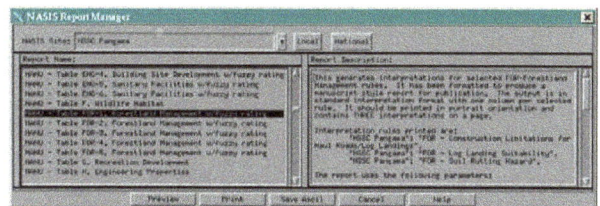

Figure 637-38   Sample Report – MANU Format

637.30 General

This subpart describes the use of the Ecological Site Information System (ESIS). ESIS is organized into two applications and associated databases – Ecological Site Description (ESD) and Ecological Site Inventory (ESI). These two applications and their associated activities are described in detail in this subpart.

Ecological Site Information is accessed via the Internet through the ESIS link on the Plants homepage at http://plants.usda.gov. Clicking on the ESIS link (see Figure 637-39) will take you to the ESIS homepage (see Figure 637-40). From the ESIS homepage you can access the Ecological Site Description application and the Ecological Site Inventory application.

Figure 637-39   Plants Homepage

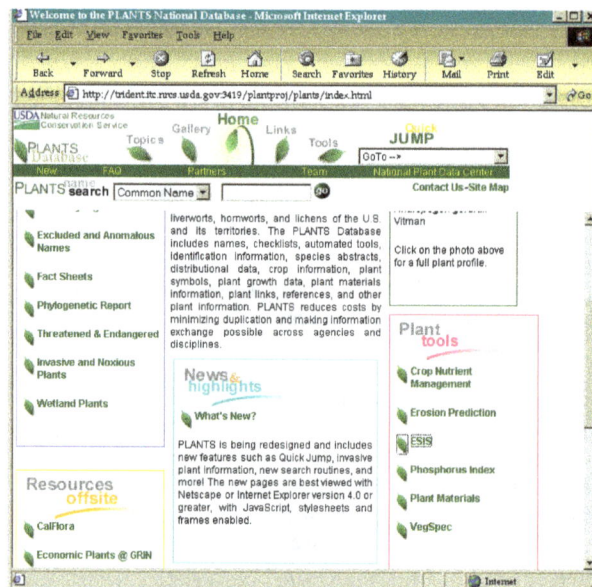

Figure 637-40   ESIS Homepage

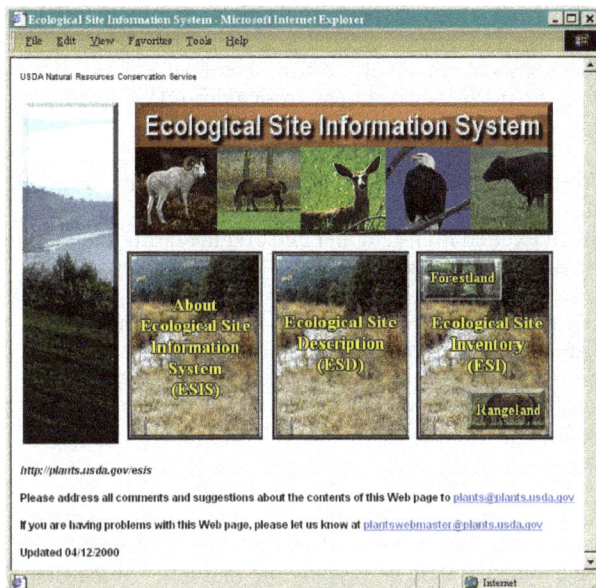

637.31 Ecological Site Description Application (ESD)

The National Forestry Manual, Part 537 provides an explanation of the ecological site concept and a detailed description of the various components comprising an ecological site description. This subpart describes the use of the Ecological Site Description (ESD) application to enter, edit, and report out the data associated with the development of ecological site descriptions.

From the ESD homepage (see Figure 637-41) you can choose to (1) enter and/or edit ecological site descriptions and (2) view reports of existing ecological site descriptions. The following subparts provide instructions in the use of these two functions.

Figure 637-41   ESD Homepage

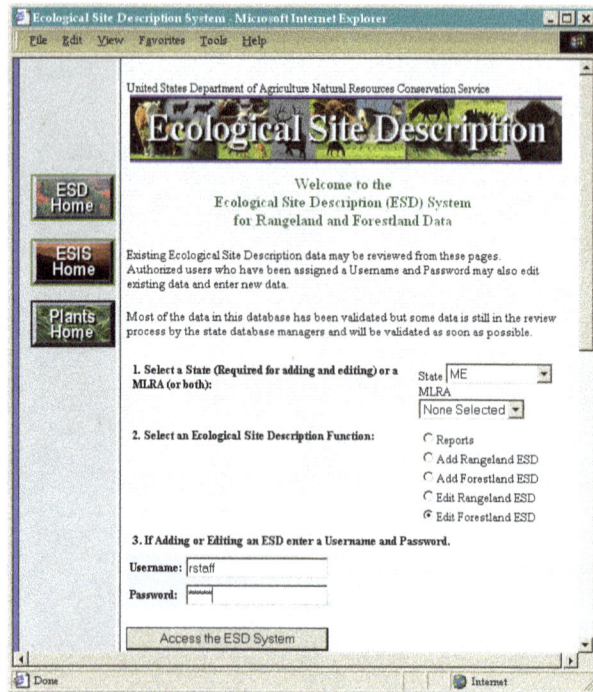

(a) Conventions for Using the ESD Application

The following conventions are used throughout the ESD application:

Text Boxes – A text box requires that you type in data. In most cases, anything can be typed in a text box, although in some cases, the data will be validated before it is accepted. The "Username" field shown in Figure 637-41 is an example of a text box.

Choice List Boxes – Choice list boxes contain a list of appropriate items from which to make a selection. Normally, only one item can be chosen. The boxes themselves vary in size and in the number of items they contain. Boxes that have more items than can be viewed in the box have a scroll bar on the right side of the box that allows you to scroll through the list of items. To select an item in a choice box, highlight the item with the mouse. Figure 637-47 and Figure 637-51 show two examples of choice lists.

Choice List Fields – Like the choice list box described above, the choice list field contains a list of appropriate items from which to make a selection. Normally, the only data that can be entered in a choice list field are data contained in the choice list. Unlike the choice list box, the list of items is view from a drop-down choice list.

To select an entry from a drop-down choice list, click on the "down arrow" located on the right edge of the choice list field, then highlight and click on the desired item in the drop-down choice list.

Some choice lists contain more items than can be displayed in the drop-down choice list. To see all of the items, use the scroll bar on the right side of the box to scroll through the items. Figure 637-42 shows an example of a drop-down list of items in a choice list field.

As an alternative to selecting from the drop-down choice list, you can also select a choice by placing the cursor in the choice list field and typing the first letter of the first word of your desired choice.

Radio Buttons – Radio buttons are a modified form of choice lists. The difference is, you are able to view all of the items on the screen, rather than having a list of items in a drop-down choice or scrollable choice list box. Radio buttons are normally used when there are only a few choices. Only one item may be selected in a list. To select an item in a radio button list, click on the round radio button. Figure 637-41 shows a radio button choice list with the item "Edit Forestland ESD" selected.

Check Boxes – Check boxes are similar to radio buttons in that they are a modified form of choice lists. Unlike radio buttons, however, more than one check box can be selected. Figure 637-53 shows and example of the use of check boxes to select Aspect.

Figure 637-42   Choice List Field

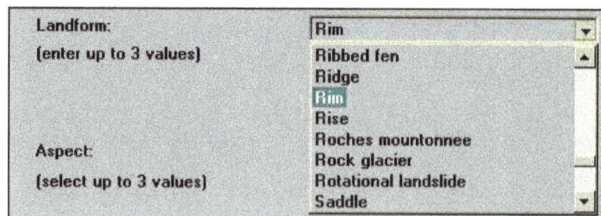

Entry/Edit Tables – Several screens utilize a table structure to display data. Figure 637-43 shows an example of a typical table. The data cannot be entered or edited directly in the table.

To **enter** data, click on the Add button at the bottom of the table and then complete the subsequent dialog box.

To **modify** previously entered data, highlight the row to be modified by clicking on the button to the left of the desired row. Then click on the Modify button at the bottom of the table and modify the data in the subsequent dialog box.

To **delete** previously entered data, highlight the row to be deleted by clicking the button to the left of the desired row. Then click on the Delete button at the bottom of the table.

Figure 637-43   Entry/Edit Tables

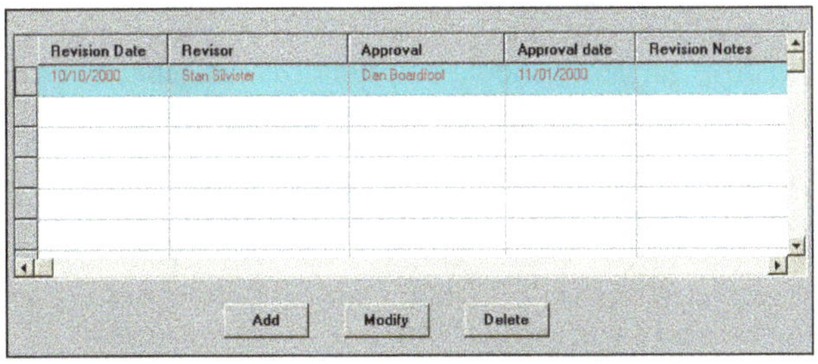

(b) Adding or Editing Ecological Site Descriptions

The ESD application is designed to allow intermittent data entry. This allows you to enter as little or as much data as are currently available to support an ecological site description. It is not required to supply all the data associated with a description at any one time. The application can be accessed to enter new data or update existing data at any time. The following describes the procedures for accessing the system to add a new ESD or edit an existing ESD.

To add or edit an ESD, complete the following on the ESD homepage (see Figure 637-41):

1. State – From the choice list, select the appropriate state. You must select a state to add or edit an ESD. Note, you can enter or edit data only for those state for which you have authorization.

2. MLRA – If you are editing an existing ESD, you may choose to select an appropriate MLRA from the

choice list, but it is not required. When editing an existing ESD, selecting an MLRA serves to reduce the number of ESD's from which to choose, if the MLRA covers only a portion of the state selected. It is not necessary to select an MLRA if you are adding a new ESD.

3. Select a Function – Click on the Add Forestland ESD radio button to add a new forestland ESD or the "Edit Forestland ESD" radio button to edit an existing forestland ESD.

4. Enter Username and Password – Enter your username and password. You cannot add or edit an ESD unless you have an authorized username and password. To obtain a username and password, or if you are experiencing difficulty with an existing username and password, contact one of the individuals listed at the bottom of the ESD homepage.

5. Access the System – Click on the Access the ESD System button.

    If you elected to add a Forestland ESD, the **Forestland (Add)** screen shown in Figure 637-44 will be displayed. To add an new ESD, follow the instructions in the **Forestland(Add) Screen** section below.

    If you elected to edit an existing Forestland ESD, the **Ecological Site Description Selection** screen shown in Figure 637-45 will be displayed. To edit an existing ESD, click on the ESD's ID number.

    Upon selecting an ESD ID, the Java applet used for editing the ESD data will be loaded. The loading of the Java applet can take several minutes, so be patient. This is especially true the first time you run the ESD application. The first time a user runs the ESD application, the program will download some rather large Java scripts to the user's computer. This will take some time (5–10 minutes). Fortunately, these files need to be downloaded only once, so subsequent access is much faster.
    At some point after clicking on the Continue button, the dialog box shown in Figure 637-48 will appear. This is normal. Just click on the OK button to continue.

    For instructions on entering data for a new ESD or

637-21

editing data for an existing ESD, refer to **Data Entry and Editing**, below.

(1) Forestland(Add) Screen
This screen is used to identify and name the site being entered (see Figure 637-44).

Site Identifier – Enter the ecological site number as outlined in the National Forestry Manual, part 637. Notice that the postal abbreviation for the state is automatically entered and cannot be edited. To change the state you must return to the ESD homepage and select another state.

Site Name – Enter the scientific name of a primary overstory species and either a primary shrub species, or a primary herbaceous species, or both. A secondary triad of overstory, shrub, and herbaceous species may also be specified but is not required. Refer to the National Forestry Manual, part 537 for details on naming forestland ESDs.

Figure 637-44   Forestland (Add) Screen

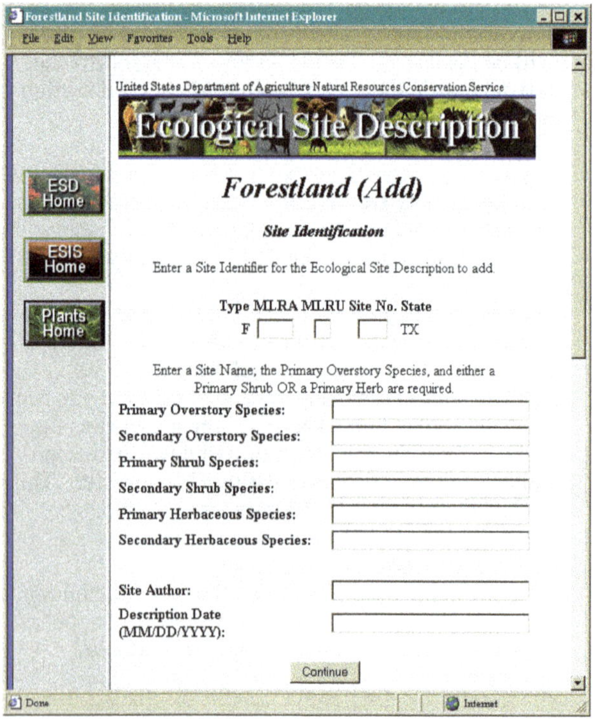

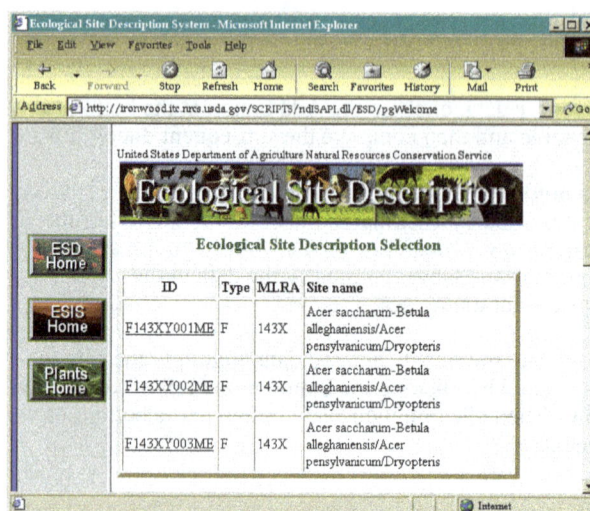

The scientific names must match those in the Plants database (http://plants.usda.gov). The names are case sensitive, so genus should be capitalized and species should be all lowercase. When you click on the Continue button, the species names are validated against the plants database. If a species name is not valid, the screen shown in Figure 637-46 will appear.

If you click on the OK button you will be returned to the **Forestland (Add)** screen where you will need to re-enter the correct scientific name.

If you click on the Plants DB button, the **Scientific Name Search** screen will be displayed (see Figure 637-47). Enter a genus name and click on the Search button to get a list of all species belonging to that genus.

Select the appropriate species and click the OK button. The **Forestland (Add)** screen is re-displayed with the selected scientific name inserted. Repeat this process until all scientific names are validated.

Figure 637-45   ESD Selection Screen

Figure 637-46   Invalid Plant Name Screen

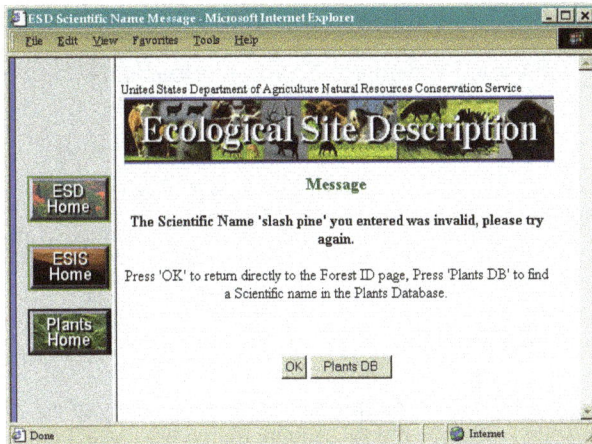

Site Author – Enter the name of the person developing the ecological site description.

Figure 637-47   Scientific Name Search Screen

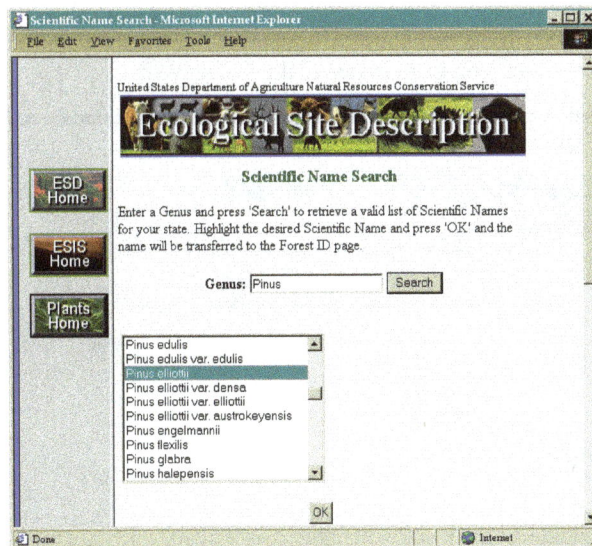

Description Date – Enter the date the description is approved for use in the MM/DD/YYYY format (i.e. 10/23/2000).

Continue – After completing data entry on the **Forestland (Add)** screen, click on the Continue button at the bottom of the screen. Validating the scientific names and starting the Java applet that is used for

entering the remainder of the site description data can take several minutes, so be patient. This is especially true the first time you run the ESD application. The first time a user runs the ESD application, the program will download some rather large Java scripts to the user's computer. This will take some time (5–10 minutes). Fortunately, these files need to be downloaded only once, so subsequent access is much faster.

At some point after clicking on the Continue button, the dialog box shown in Figure 637-48 will appear. This is normal. Just click on the OK button to continue.

For instructions on entering data for a new ESD or editing data for an existing ESD, refer to **Data Entry and Editing**, below.

Figure 637-48   Applet Initialized Dialog Box

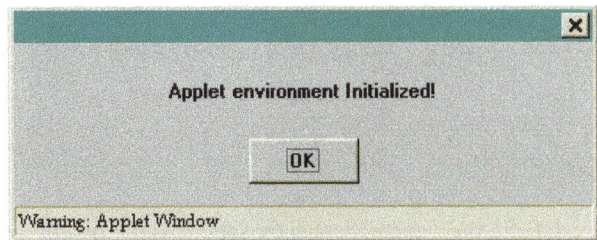

(2) Data Entry and Editing
After completing the requirements for adding a new ESD or after selection of an existing ESD, as described above, you are ready to begin entering data for a new ESD or editing data for an existing ESD.

Data entry in the ESD application is organized into eight data entry sections:

- General
- Physiographic Features
- Climate Features
- Water Features
- Soil Features
- Plant Communities
- Site Interpretations
- Supporting Information

As shown in Figure 637-49, the frame on the right side of the screen contains the data entry fields associated with one of the eight data entry sections and the frame on the left side of the screen contains a list of the eight data entry sections. You can jump to the different sections by clicking on the section names. The screen associated with the section name is displayed on the

637-23

right side while the section list on the left side remains visible.

Figure 637-49   Forest Site General Information Screen

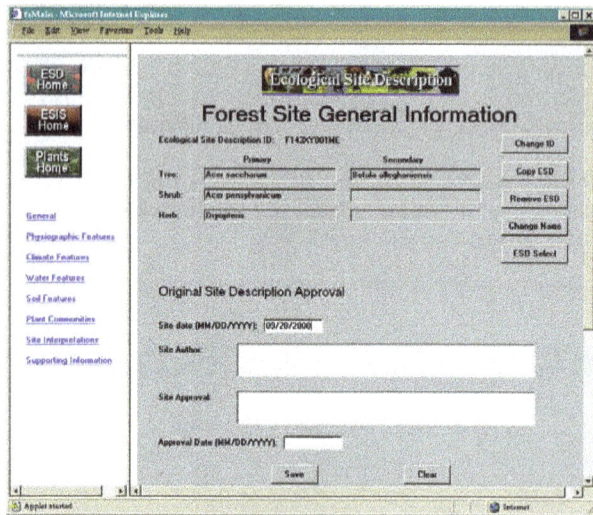

By default, the General data entry section is the first to be displayed after completing the requirements for adding a new ESD or selecting an existing ESD. However, you can complete the sections in any order desired.

The following subparts provide detailed instructions for entering or modifying data in the eight data entry sections.

(i) General Section
This section provides tools to manage the basic information about an ecological site. Figure 637-49 shows and example of the screen associated with this section.

Change ID – Click on this button to change the ESD ID. On the subsequent dialog box, enter the new ESD ID and click the OK button (see Figure 637-50).

Figure 637-50   Change ESD ID

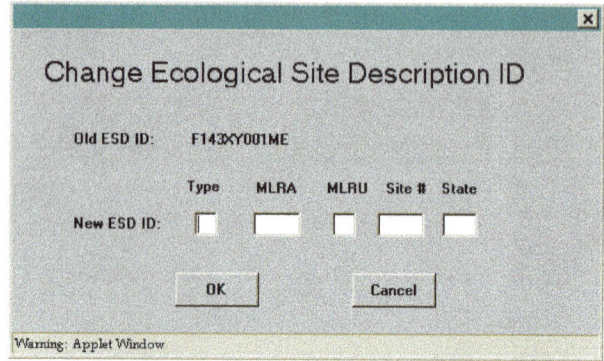

Copy ESD – Click on this button to make a copy of an existing ESD. On the subsequent dialog box, select the ESD ID that you want to copy, enter a new ESD ID, and click the OK button (see Figure 637-51). Note, you must assign a unique ID to the new site. The ability to copy an existing ESD could save considerable data entry time.

Remove ESD – Click on this button to delete an entire ESD from the database. A warning will be displayed to verify that you want to delete the ESD.

Figure 637-51   Copy ESD

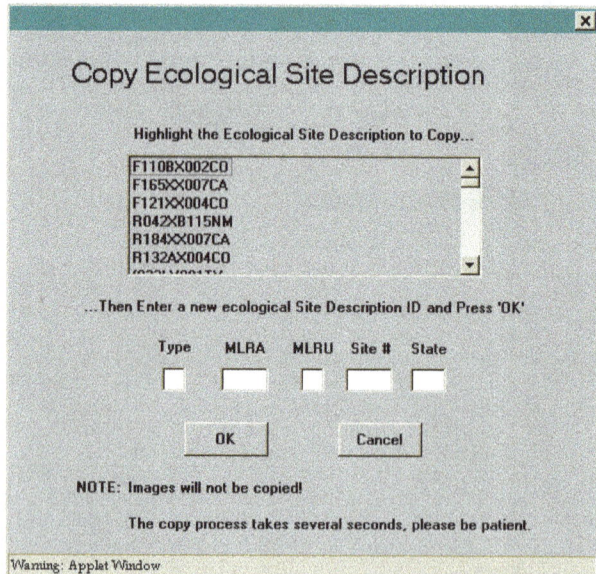

Change Name – Click on this button to change all or any part of the ESD name. On the subsequent dialog box (see Figure 637-52), enter new primary or secondary species, or modify the current ones and click on Save. Click on the Find… button for assistance in identifying species.

Figure 637-52   Change Name

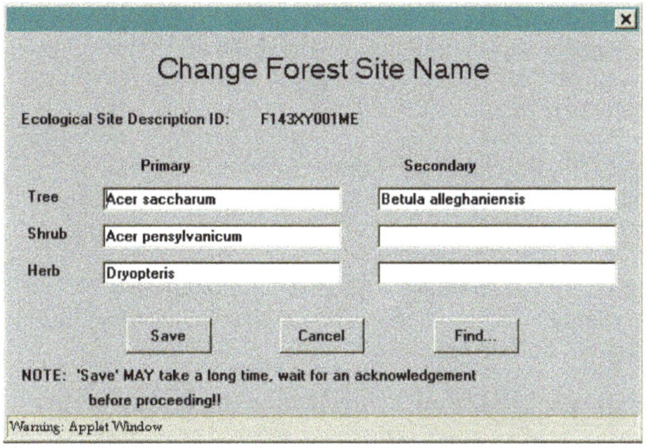

ESD Select – Click on this button to change to another ESD. This function is useful if you want to edit several different sites without having to re-logon each time you change sites. From the subsequent screen, select the desired site.

Site Date – Enter the date the site is developed, either partially or completely.

Site Author – Type in the name of the individual responsible for initial development of the ESD.

Site Approval – Type in the name of the individual responsible for approving ESD's in your area.

Approval Date – Type in the date the site is approved for use.

Revision Table – The table at the bottom of the screen is used to record information relative to site revisions. Whenever a site is revised after the initial approval date, record the revision date, reviser's name, approver's name, approval date, and any notes pertinent to the revision.

After completing the **General** section, click on the Save button to save the data to the database.

(ii) Physiographic Features Section
This section is used to record the physiographic features of an ecological site. Remember, these data should reflect the range of values representative of the various soil map unit components that comprise the site. Figure 637-53 shows an example of the screen associated with this section.

Figure 637-53   Physiographic Features Screen

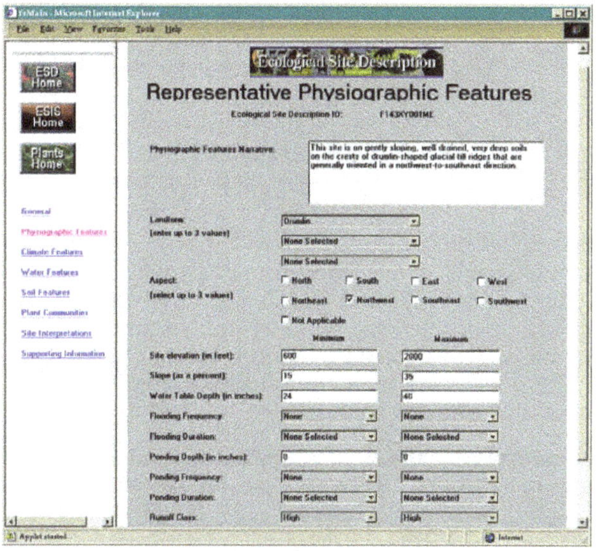

Physiographic Features Narrative – Enter a narrative description of the physiographic features of the site.

Landform – Select up to three landforms from the choice list.

Aspect – Select up to three aspects by selecting each check box that applies.

Elevation – Enter the minimum and maximum elevation of the site, in feet.

Slope – Enter the minimum and maximum slope of the site, in percent.

Water Table Depth – Enter the minimum and maximum depth to water table for the site, in inches.

Flooding Frequency – Select the minimum and maximum flooding frequency for the site from the choice lists.

Flooding Duration – Select the minimum and maximum flooding duration for the site from the choice lists.

Ponding Depth – Enter the minimum and maximum ponding depth for the site.

Ponding Frequency – Select the minimum and maximum ponding frequency for the site from the choice lists.

Ponding Duration – Select the minimum and maximum ponding duration for the site from the choice lists.

Runoff Class – Select the minimum and maximum runoff class designation for the site from the choice lists.

After completing the **Physiographic** section, click on the Save button to save the data to the database.

(iii) Climate Features Section
This section is used to record the representative climate features of an ecological site. Remember, these data should reflect the range of values representative of the various soil map unit components that comprise the site. Figure 637-54 shows an example of the screen associated with this section.

Figure 637-54   Climate Features Screen

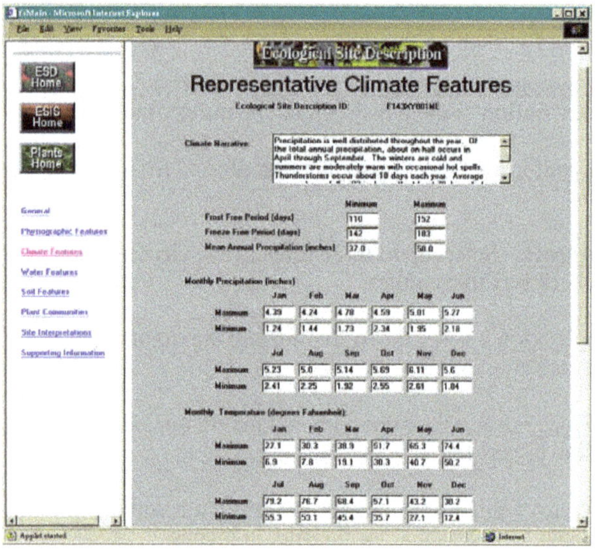

Climate Narrative – Enter a narrative description of the climatic features of the site.

Frost Free Period – Enter the minimum and maximum number of days the site is free of frost.

Freeze Free Period – Enter the minimum and maximum number of days the site does not have freezing temperatures.

Mean Annual Precipitation – Enter the minimum and maximum average precipitation the site receives annually, in inches.

Monthly Precipitation – Enter the minimum and maximum average precipitation the site receives each month, in inches.

Monthly Temperature – Enter the minimum and maximum average temperature at the site each month, in degrees Fahrenheit.

Climate Station Table – The table at the bottom of the screen is used to record the unique identifier of the climate station(s) from which the climate data was derived. Enter the climate station ID, climate station name, and the start and end dates used to derive the climate averages.

After completing the **Climate Features** section, click on the Save button to save the data to the database.

(iv) Water Features Section
This section is used to record the representative water features that influence the plant communities on the site. Remember, these data should reflect the range of values representative of the various soil map unit components that comprise the site. Figure 637-55 shows an example of the screen associated with this section.

Water Features Narrative – Enter a narrative description of the influencing water features on the site. If the plant communities are not influenced by water from a wetland or stream, then state this in the narrative.

Wetland Description Table – The table at the bottom of the screen is used to record the wetland systems and stream types (if applicable) that influence plant communities on the site.

Figure 637-56 shows and example of the data entry screen associated with the wetland description table. On this screen select the applicable wetland system, wetland subsytem, and wetland class from the choice lists.

Figure 637-55   Influencing Water Features Screen

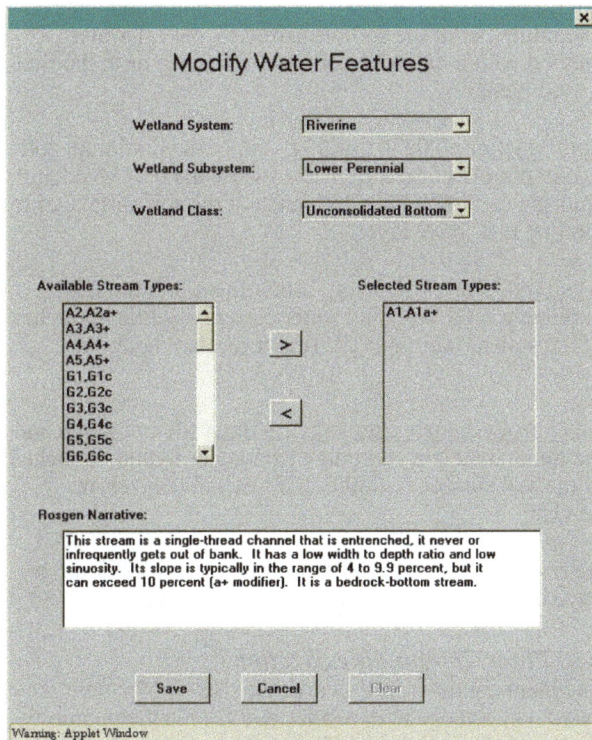

Figure 637-56   Water Features Screen

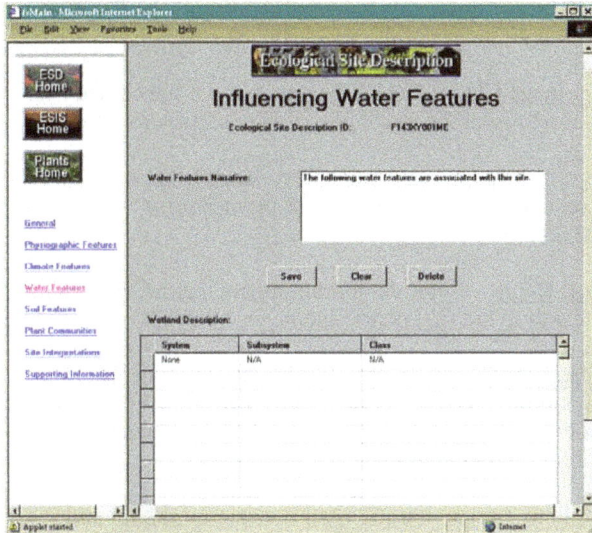

Stream types can be selected only when a wetland system of Riverine is selected. To select the stream type(s) associated with a riverine wetland system, click on the applicable stream type(s) in the choice list box and then click on the > arrow. The stream type(s) selected will appear in the Selected Stream Types box and the narrative associated with the stream type will appear in the Rosgen Narrative box.

To "unselect" a stream type, click on the applicable stream type in the Selected Stream Types box and then click on the < arrow.

Click on Save at the bottom of the water features data entry screen to save the data and return to the **Influencing Water Features** screen.

After completing the **Influencing Water Features** section, click on the Save button to save the data to the database.

(v) Soil Features Section
This section is used to record the representative soil features for the site. Remember, these data should reflect the range of values representative of the various soil map unit components that comprise the site. Figure 637-57 shows an example of the screen associated with this section.

Soil Features Narrative – Enter a narrative description of the soil features of the site.

Parent Material Kind – Select the predominant kind of parent material from the choice list.

Parent Material Origin – Select the origin of the selected kind of parent material from the choice list.

Surface Texture – Select up to three texture classes for the surface of the soil from the choice list. The depth of the soil surface represented by the selected texture classes should be recorded in the Soil Features Narrative, e.g. "The surface textures recorded represent those found within 8 inches of the soil surface."

Surface Texture Modifier – If applicable, select a texture modifier from the choice list. The texture modifier for each texture class should be selected in the choice list field directly below the choice list field of the texture class it modifies.

637-27

Figure 637-57   **Representative Soil Features Screen**

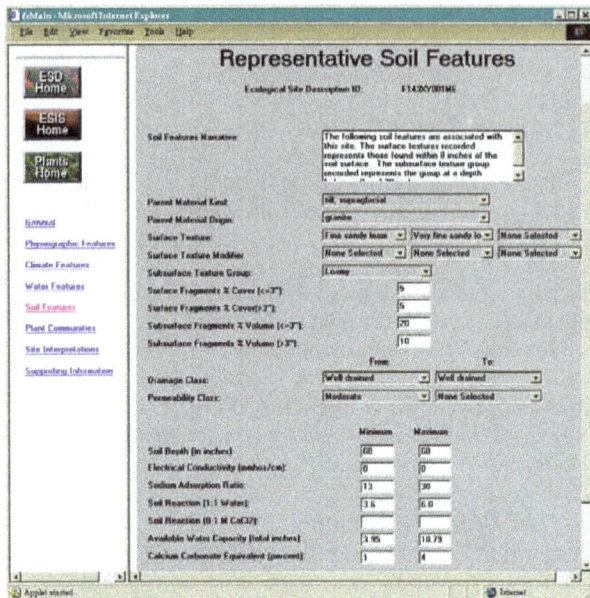

Subsurface Texture Group – Select the general term that describes the texture class of the soil below the soil surface. The depth range for which the texture group applies should be recorded in the Soil Features Narrative, e.g. "The subsurface texture group recorded represents the group at a depth between 8 and 20 inches." Refer to Part 537.31 of the National Forestry Manual for texture classes to assign to each texture group.

Surface Fragments <=3% – Enter the percent of the soil surface covered by rock fragments less than or equal to 3" in size.

Surface Fragments >3" – Enter the percent of the soil surface covered by rock fragments greater than 3" in size.

Subsurface Fragments <=3" – Enter the percent, by volume, of the rock fragments less than or equal to 3" in size within the soil profile to a specified depth. The depth range for which subsurface fragments apply should be recorded in the Soil Features Narrative, e.g. "The subsurface fragments recorded represent the fragments within 40 inches of the soil surface."

Subsurface Fragments >3" – Enter the percent by, volume, of the rock fragments greater than 3" in size

within the soil profile to a specified depth. The depth range for which subsurface fragments apply should be recorded in the Soil Features Narrative, e.g. "The subsurface fragments recorded represents the fragments within 40 inches of the soil surface".

Drainage Class – Select the values from the choice lists that represent the range of drainage classes for the site.

Permeability Class – Select the values from the choice lists that represent the range of permeability classes for the site.

Soil Depth – Enter the minimum and maximum depth of the soil to the first restrictive layer, in inches.

Electrical Conductivity – Enter the minimum and maximum electrical conductivity values within 40 inches of the soil surface or to the first restrictive layer, in millimhos per centimeter.

Sodium Adsorption Ratio – Enter the minimum and maximum sodium adsorption ratio values within 40 inches of the soil surface or to the first restrictive layer.

Soil Reaction (1:1 Water) – Enter the minimum and maximum pH of the soil measured by the 1:1 water method within 40 inches of the soil surface or to the first restrictive layer.

Soil Reaction (0.01M CaC12) – Enter the minimum and maximum pH of the soil measured by the 0.01M calcium chloride method within 40 inches of the soil surface or to the first restrictive layer.

Available Water Capacity – Enter the minimum and maximum total available water capacity within 40 inches of the soil surface or to the first restrictive layer, in inches.

Calcium Carbonate Equivalent – Enter the minimum and maximum calcium carbonate equivalent within 40 inches of the soil surface or to the first restrictive layer, in percent.

After completing the **Soil Features** section, click on the Save button to save the data to the database.

(vi) Plant Communities Section
The plant communities section consists of two main parts: (1) ecological dynamics and (2) plant community. Each of these is described below.

(1) Forest Plant Communities Screen – Ecological Dynamics

This portion of the **Forest Plant Communities** screen is used to record the ecological dynamics of the site (see Figure 637-58).

Figure 637-58   Plant Communities Screen -- Ecological Dynamics Portion

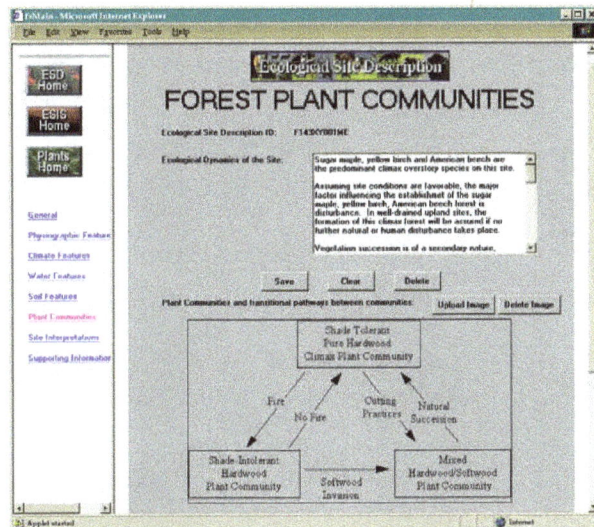

Ecological Dynamics of the Site – Enter a narrative description of the ecological dynamics of the site.

Save – Click on this button to save the current text in the Ecological Dynamics of the Site text box.

Clear – Click on this button to clear the current text in the Ecological Dynamics of the Site text box.

Delete – Click on this button to delete the current text in the Ecological Dynamics of the Site text box from the database.

Upload Image – Click on this button to upload an image of a state and transition model that shows the vegetation states and the transition pathway between the states. Complete the information on the subsequent Add An Image dialog box (see Figure 637-59), and then click on the Add Image button to upload the image. The image should appear when you return to the **Forest Plant Communities** screen.

NOTE: All image files must be of the file type *JPEG* or *GIF* and cannot be larger than 250KB in size.

Delete Image – Click on this button to delete the state and transition model image from the database.

Figure 637-59   Add An Image Dialog Box

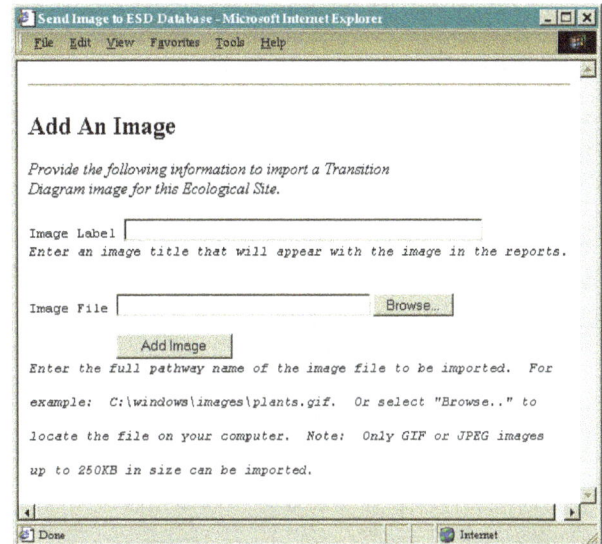

(2) Forest Plant Communities Screen – Plant Community

This part of the **Forest Plant Communities** screen displays the various vegetative states that are associated with the ESD (see Figure 637-60). These vegetative states should correspond to those shown in the transition pathways diagram.

Figure 637-60   Forest Plant Communities Screen – Plant Community Portion

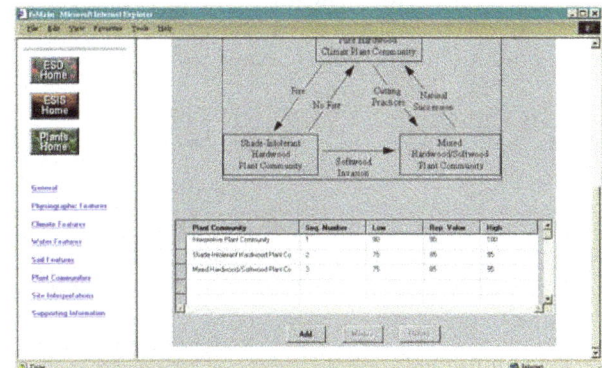

637-29

Add – Click on the Add button at the bottom of the Plant Community table to add a new vegetative state. The screen shown in Figure 637-61 will be displayed.

Plant Community Name – Enter a descriptive name for the vegetative state. In most cases, the name will be the same as that shown in the transition pathways diagram for the corresponding vegetation state.

Figure 637-61   Add Forest Plant Community

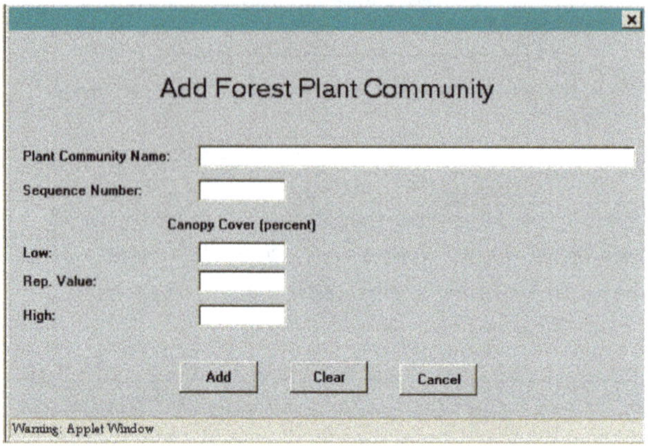

Plant Community Sequence Number – Enter the sequence number for the vegetative state, such as 1 for the historic plant community. This sequence number is not critical at this time, but in future it may be used to automatically generate a transition diagram.

Low Understory Canopy Cover (%) – Enter the percent of canopy cover that would represent the minimum canopy cover for the vegetative state.

Rep. Value Understory Canopy Cover (%) – Enter the percent of canopy cover that would represent the typical canopy cover for the vegetative state.

High Understory Canopy Cover (%) – Enter the percent of canopy cover that would represent the maximum canopy cover for the vegetative state.

Add – Click on the Add button to save the entered data. After the data have been saved, the **Forest Plant Community** screen will be displayed (see Figure 637-62).

Modify – Highlight a row in the Plant Community table and click on the Modify button at the bottom of the table to modify a previously entered vegetative state. The

Forest Plant Community screen will be displayed (see Figure 637-62).

The **Forest Plant Community** screen is used to enter or edit data associated with each vegetative state. The screen consists of four main portions. Each of its parts is detailed below.

Figure 637-62   Forest Plant Community Screen – Ecological Site ID Portion

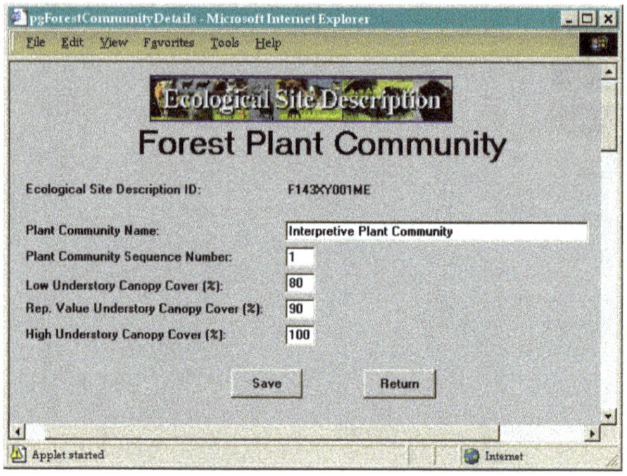

The part of the **Forest Plant Community** shown in Figure 637-62 screen is used to record the name of the vegetative state and the typical canopy cover percentages. It contains the following items.

Ecological Site Description ID – This field is read-only and displays the ESD ID of the current site.

Plant Community Name – Enter the descriptive name for the vegetative state. In most cases, the name should be the same as that shown in the transition pathways diagram for the corresponding vegetation state.

Plant Community Sequence Number – Enter the sequence number for the vegetative state, such as 1 for the historic plant community. This sequence number is not critical at this time, but in future it may be used to automatically generate a transition diagram.

Low Understory Canopy Cover (%) – Enter the percent of canopy cover that would represent the minimum canopy cover for the vegetative state.

637-30

Rep. Value Understory Canopy Cover (%) – Enter the percent of canopy cover that would represent the typical canopy cover for the vegetative state.

High Understory Canopy Cover (%) – Enter the percent of canopy cover that would represent the maximum canopy cover for the vegetative state.

Save – Click on this button to save the information entered in the above data fields.

Return – Click on this button to return to the **Forest Plant Communities Screen**. Be sure to save any data entered or edited before clicking on the return button.

Figure 637-63   Forest Plant Community Screen – Plant Narrative Portion

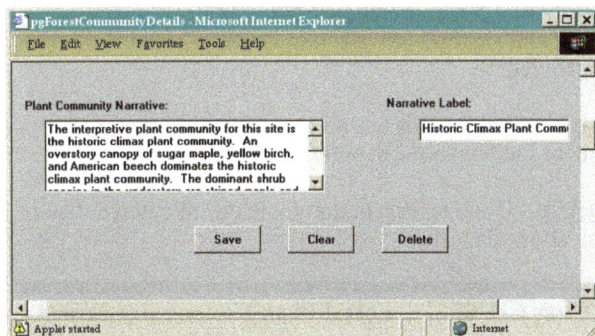

The part of the **Forest Plant Community** screen shown in Figure 637-63 is used to record a narrative description of a vegetative state. It contains the following items.

Plant Community Narrative – Enter a narrative description of the vegetative state. Where the vegetative state represents the interpretive plant community for the site, the first sentence in narrative should clearly state whether the interpretive plant community described is the historic climax or the naturalized plant community.

Narrative Label – Enter a label for the vegetative state. In most cases the label will be the same as the plant community name, but not always. For example, the plant community may be named "Interpretive Plant Community," but the label may be "Historic Climax Community" or "Naturalized Plant Community."

Save – Click on this button to save the information entered in the Plant Community Narrative and Narrative Label fields.

Clear – Click on this button to clear all entered data in the Plant Community Narrative and Narrative Label fields.

Delete – Click on this button to delete all entered data in the Plant Community Narrative and Narrative Label fields.

Figure 637-64   Forest Plant Community Screen – Ground Cover and Structure Table Portion

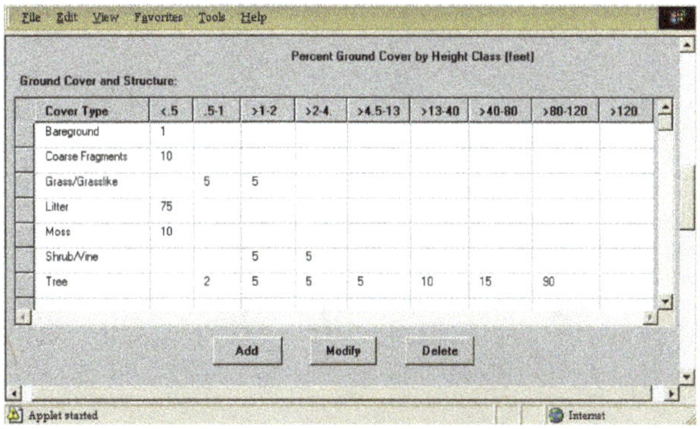

The part of the **Forest Plant Community** screen shown in Figure 637-64 displays the percent ground cover by height class for various cover types typical of the vegetative state being described. It contains the following items.

Add/Modify – Click on the Add or Modify button at the bottom of the Ground Cover and Structure table to record new or modify existing ground cover and structure data. For new entries the screen shown in Figure 637-65 will be displayed. When modifying previously entered data, a screen identical to the screen shown in Figure 637-65 will be displayed, except the screen is titled Modify rather than Add. Refer to the National Forestry Manual, Part 537 for details on recording ground cover and structure.

Record the percent cover by height class for each cover type typical of the vegetative state. After recording the percent cover for a cover type, click on the Save button to insert the data into the database, then enter the next percent cover for the next cover type. When the percent cover for all cover types has been entered, click on the Return button to return to the **Forest Plant Community**

637-31

screen. Click on the Clear button to clear all entries. The Clear button will not delete previously saved entries.

Figure 637-65   Forest Cover and Structure

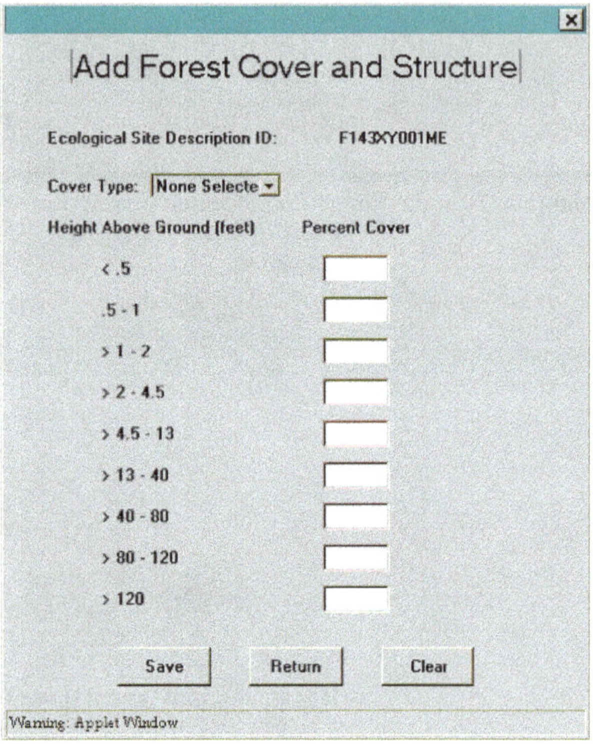

Delete – Highlight a row in the Ground Cover and Structure table and click on the Delete button at the bottom of the table to delete all data for that particular row from the database.

Figure 637-66   Forest Plant Community Screen – Overstory/Understory Narrative Portion

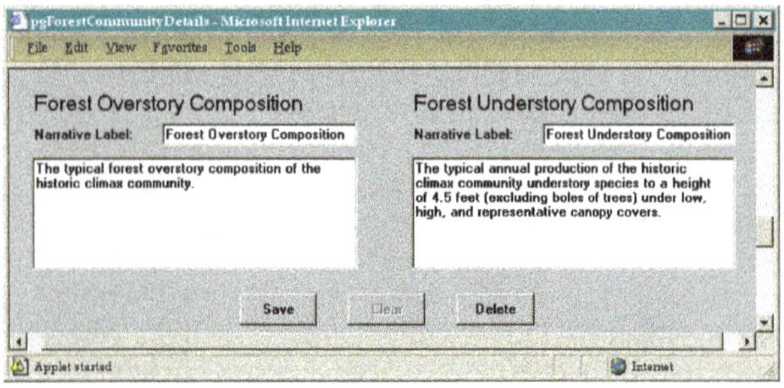

The part of the **Forest Plant Community** screen shown in Figure 637-66 is used to record a narrative description of the forest overstory and understory composition. It contains the following items.

Narrative Labels – Record a label for the overstory and understory narratives. These labels will be used as headers for the narratives in the ESD report.

Forest Overstory Composition Narrative – Record a narrative to be used as an introduction to the percent overstory composition table in the ESD report.

Forest Understory Composition Narrative – Record a narrative to be used as an introduction to the percent understory composition table in the ESD report. By default, the narrative "The typical annual production of the historic climax community understory species to a height of 4.5 feet (excluding boles of trees) under low, high, and representative canopy covers" will be displayed.

Save – Click on this button to save the information entered in the above data fields.

Delete – Click on this button to delete all entered data in the above data fields.

Figure 637-67   Forest Plant Community Screen – Composition and Production Table Portion

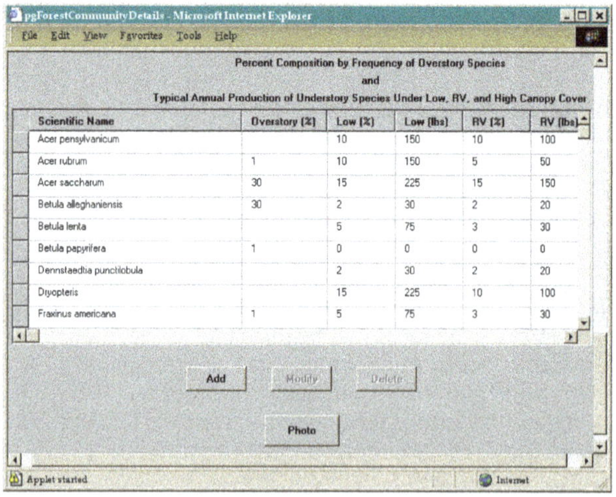

The part of the **Forest Plant Community** screen shown in Figure 637-67 displays the percent composition of the overstory species and the annual production of understory species. It contains the following items.

Add/Modify – Click on the Add or Modify button at the bottom of the table to record new or modify existing data. For new entries the screen shown in Figure 637-68 will be displayed. When modifying previously entered data, a screen identical to the screen shown in Figure 637-68 will be displayed, except the screen is titled Modify rather than Add.

Figure 637-68   Forest Overstory/Understory

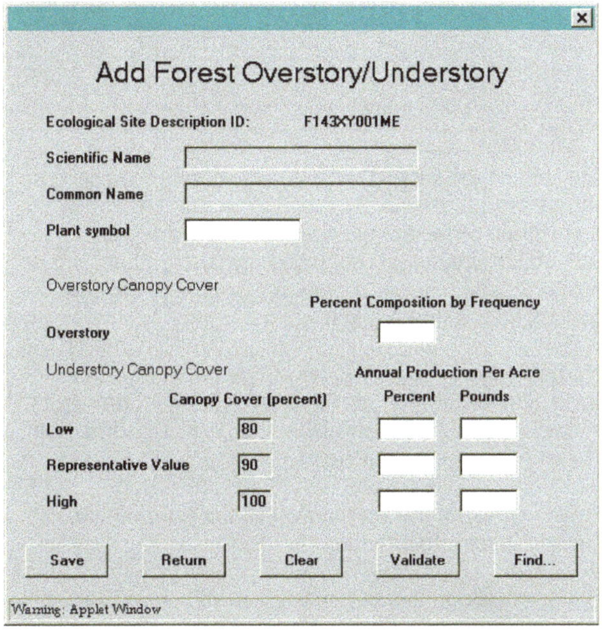

For each overstory or understory species to be recorded, enter the plant symbol and click on the Validate button. If the plant symbol entered is valid, the scientific name and common name will be displayed. To look up a plant symbol, click on the Find button. The screen shown in Figure 637-69 will be displayed. To look up a plant symbol, enter a whole or partial plant symbol in the Search Pattern field and click on the Find button. For example, to search for all plant symbols that begin with the letters PI, enter PI%. From the subsequent list, highlight the desired plant symbol and click on the Select button to automatically populate the appropriate fields on **the Forest Overstory/Understory** screen.

After selection of a valid plant symbol, enter the appropriate data in the Overstory Percent Composition by Frequency and/or Annual Production fields. The low, representative, and high canopy cover percentages

displayed are those entered on a previous screen (see Figure 637-62) and are not editable on this screen.

After entering the appropriate data, click on the Save button to save the data to the database, then continue entering data for other species, if appropriate. When data for all species have been entered, click on the Return button to return to the **Forest Plant Community** screen. Click on the Clear button to clear all entries. The Clear button will not delete previously saved entries.

Figure 637-69   Find Plant Symbol

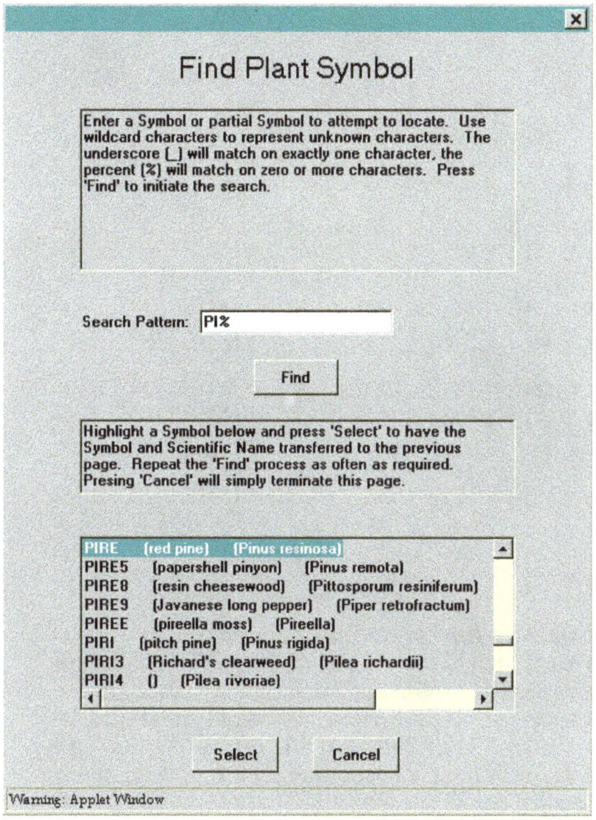

Delete – Highlight a row in the Composition and Production Table and click on the Delete button at the bottom of the table to delete all data for that particular row from the database.

Photo – Click on the Photo button to upload a photograph which typifies the vegetative state. Clicking on the Photo button will display the screen shown in Figure 637-70. To upload an image, click on the Upload Image button and complete the information on the subsequent Add An Image dialog box (see Figure 637-59), and then click on the Add Image button to upload the image. To delete a previously uploaded image, click on the Delete Image button. Click on the OK button to dismiss the screen and return to the **Forest Plant Community** screen.

NOTE: All image files must be of the file type *JPEG* or *GIF* and cannot be larger than 250KB in size. In order to display properly in the ESD report, photographic images should not exceed 500 by 500 pixels in size.

Figure 637-70   Typical Plant Community Photo

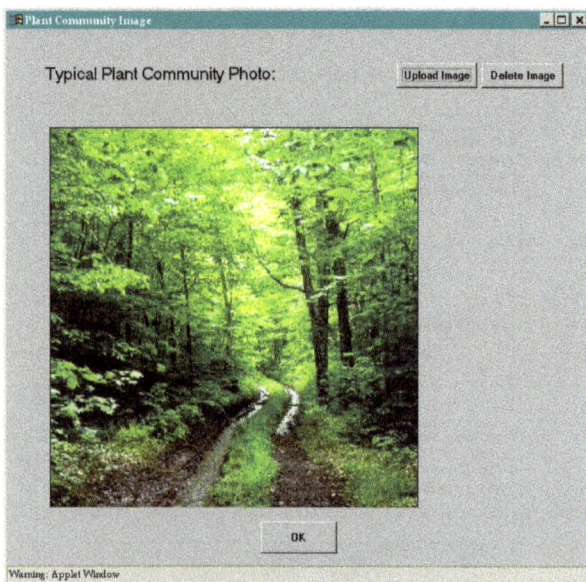

(vii) Site Interpretations Section
This section of the ESD application is used to record interpretive information pertinent to the use and management of the ecological site and its related resources. The **Ecological Site Interpretations** screen consists of three main portions. Each of these portions is detailed below.

Figure 637-71   Ecological Site Interpretations Screen – Narrative Portion

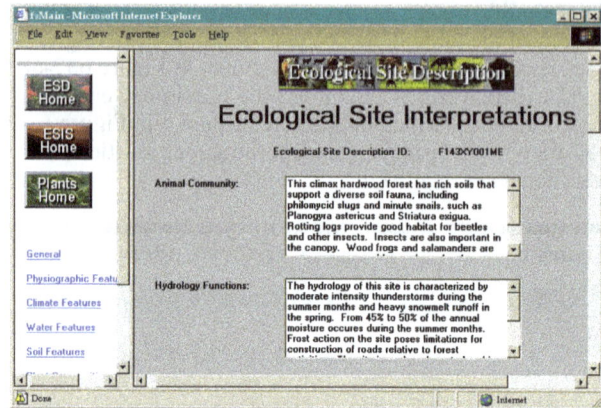

The part of the **Ecological Site Interpretations** screen shown in Figure 637-71 is used to record a narrative description of the animal community, hydrology function, recreational use, wood products, other products, and other information pertinent to the site. It contains the following items.

Narrative Text Fields – Enter the pertinent information applicable to the site. Refer to the National Forestry Manual, Part 537 for details on the type of information to record in these narratives.

Save – Click on this button to save the information entered in narrative fields.

Clear – Click on this button to clear all entered data in all the narrative fields.

Delete – Click on this button to delete the data in all the narrative fields.

Figure 637-72   **Ecological Site Interpretations Screen – Plant Preference by Animal Kind Portion**

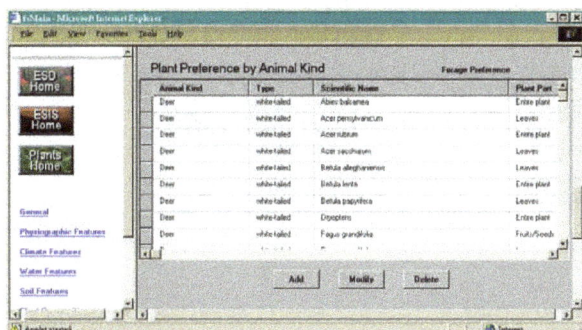

The part of the **Ecological Site Interpretations** screen shown in Figure 637-72 displays the Plant Preference by Animal Kind table that lists forage preference ratings for various plant species by plant part and time of year. It contains the following items.

Add/Modify – Click on the Add or Modify button at the bottom of the table to record new or modify existing data. For new entries, the screen shown in Figure 637-73 will be displayed. When modifying previously entered data, a screen identical to the screen shown in Figure 637-73 will be displayed, except the screen is titled Modify rather than Add.

Select the applicable animal kind from the choice list and record the forage preference by month for the applicable plant parts for each plant species. The animal type field is used to further identify animal species. For example, if "deer" is selected from the animal kind choice list, you could further identify the animal type as "mule" in the animal type field. The choice list for the plant scientific name field contains a list of all plant species previously entered for the site.

After entering the appropriate data, click on the Save button to save the data to the database, then continue entering data for other animal types, plant species, and plant parts, if appropriate. When all data have been entered, click on the Return button to return to the **Forest Plant Community** screen. Click on the Clear button to clear all entries. The Clear button will not delete previously saved entries.

Figure 637-73   **Animal Plant Preferences Screen**

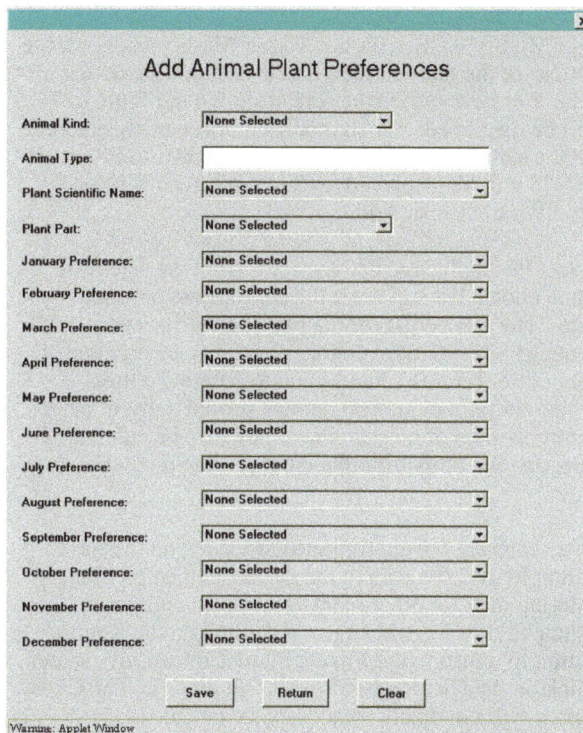

Delete – Highlight a row in the Plant Preference by Animal Kind table and click on the Delete button at the bottom of the table to delete all data for that particular row from the database.

Figure 637-74   **Ecological Site Interpretations Screen – Forest Productivity Portion**

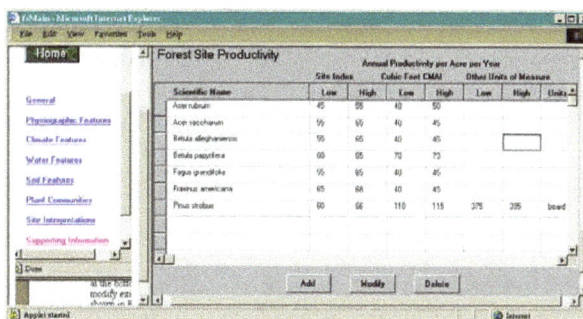

The part of the **Ecological Site Interpretations** screen shown in Figure 637-74 displays the Forage Production table that lists the site index and annual productivity of

637-35

the major tree species on the site. It contains the following items.

Add/Modify – Click on the Add or Modify button at the bottom of the table to record new or modify existing data. For new entries the screen shown in Figure 637-75 will be displayed. When modifying previously entered data, a screen identical to the screen shown in Figure 637-75 will be displayed, except the screen is titled Modify rather than Add.

Select the applicable plant species from the scientific name choice list and enter the appropriate productivity data. The choice list for the plant scientific name field contains a list of all overstory plant species previously entered for the site. In addition to entering annual productivity in cubic feet, annual productivity in other common units of measurement may also be entered by selecting the units from the other productivity choice list.

After entering the appropriate data, click on the Save button to save the data to the database, then continue entering data for other plant species, if appropriate. When all data have been entered, click on the Return button to return to the **Forest Plant Community** screen. Click on the Clear button to clear all entries. The Clear button will not delete previously saved entries.

Delete – Highlight a row in the Forest Site Productivity table and click on the Delete button at the bottom of the table to delete all data for that particular row from the database.

Figure 637-75   Forest Site Productivity Screen

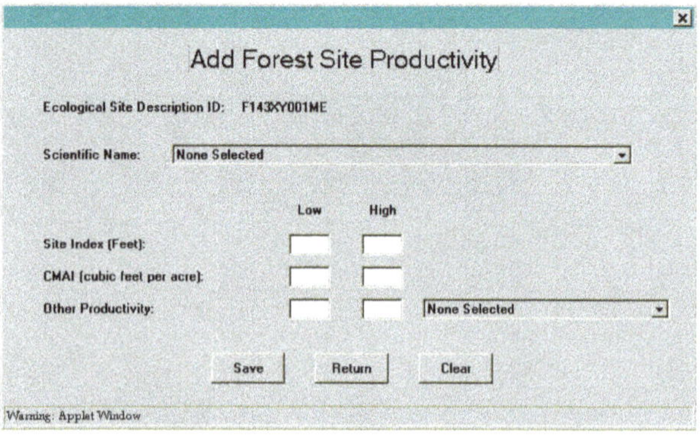

(viii) Supporting Information Section
This section of the ESD application is used to record information useful in assessing the quality of the site description and its relationship to other ecological sites.

The **Supporting Information** screen consists of seven parts. Each of these is detailed below.

Figure 637-76   Supporting Information Screen – Associated Sites Portion

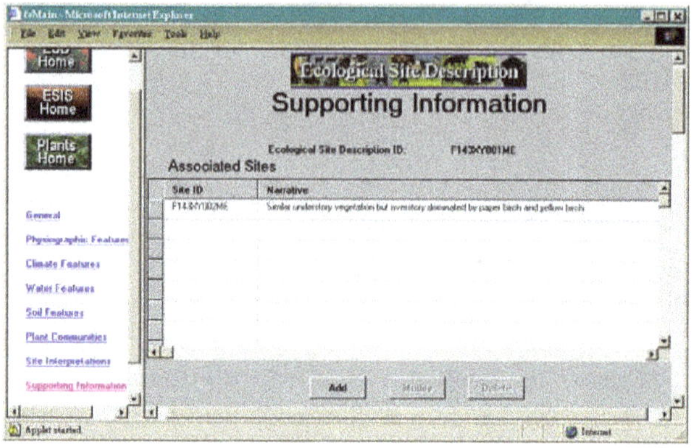

The part of the **Supporting Information** screen shown in Figure 637-76 displays the Associated Sites table that lists sites that are commonly located in conjunction with the site being described. It contains the following items.

Add/Modify – Click on the Add or Modify button at the bottom of the table to record new or modify existing data. For new entries, the screen shown in Figure 637-77 will be displayed. When modifying previously entered data, a screen identical to the screen shown in Figure 637-77 will be displayed, except the screen is titled Modify rather than Add.

Select an associated site from the Select an Existing Site choice list or enter a site by entering the appropriate data in the Input a New Site fields. A narrative describing similarities to and differences from the site being described may also be entered in the narrative field.

After entering the appropriate data, click on the Save button to save the data to the database, then continue entering other associated sites, if appropriate. When all data have been entered, click on the Return button to return to the **Supporting Information** screen. Click on the Clear button to clear all entries. The Clear button will not delete previously saved entries.

Delete – Highlight a row in Associated Sites table and click on the Delete button at the bottom of the table to delete all data for that particular row from the database.

637-36

Figure 637-77   Associated Site Screen

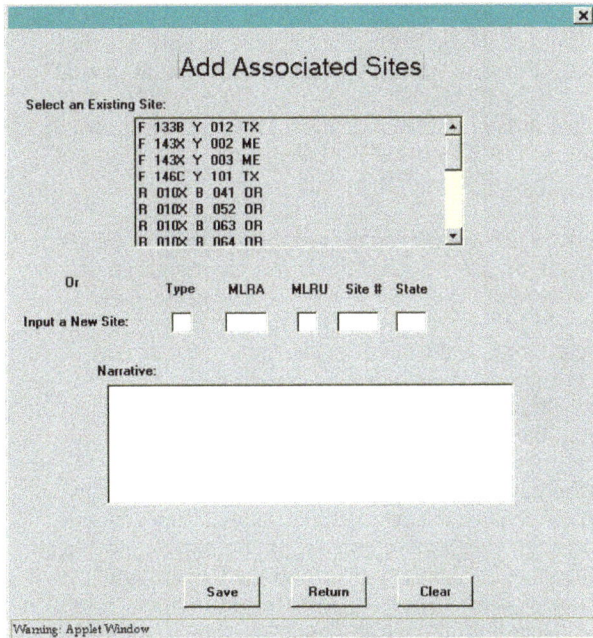

**Figure 637-78   Supporting Information Screen –
Similar Sites Portion**

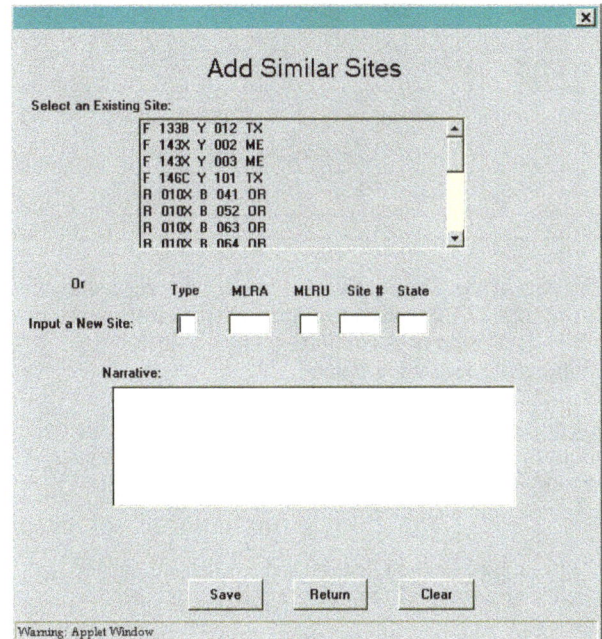

The part of the **Supporting Information** screen shown in Figure 637-78 displays the Similar Sites table that lists sited that resemble or can be confused with the site being described. It contains the following items.

Add/Modify – Click on the Add or Modify button at the bottom of the table to record new or modify existing data. For new entries, the screen shown in Figure 637-79 will be displayed. When modifying previously entered data, a screen identical to the screen shown in

Figure 637-79 will be displayed, except the screen is titled Modify rather than Add.

Figure 637-79   Similar Site Screen

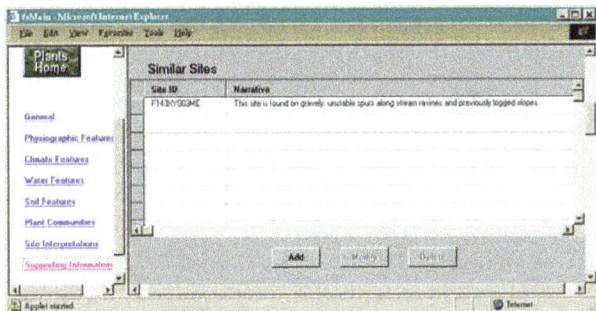

Select a similar site from the Select an Existing Site choice list or enter a site by entering the appropriate data in the Input a New Site fields. A narrative describing similarities to and differences from the site being described may also be entered in the narrative field.

After entering the appropriate data, click on the Save button to save the data to the database, then continue entering other similar sites, if appropriate. When all data have been entered, click on the Return button to return to the **Supporting Information** screen. Click on the Clear button to clear all entries. The Clear button will not delete previously saved entries.

Delete – Highlight a row in the Similar Sites table and click on the Delete button at the bottom of the table to delete all data for that particular row from the database.

637-37

**Figure 637-80   Supporting Information Screen –
Inventory Data References
Narrative Portion**

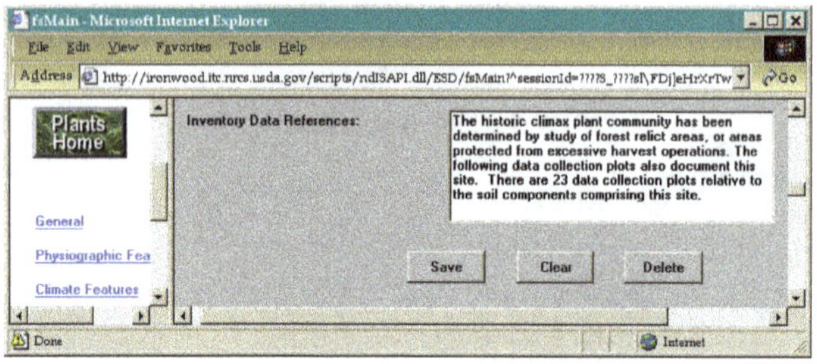

The part of the **Supporting Information** screen shown in Figure 637-80 is used to enter a narrative description of the data references supporting the site description. It contains the following items.

Narrative – Enter a narrative description of how the data about the interpretive plant community were obtained and any other pertinent information relative to site inventory plots supporting the site description.

Clear – Click on the Clear button to clear all entries. The Clear button will not delete previously saved entries.

Delete – Click on the Delete button to delete the narrative from the database.

**Figure 637-81   Supporting Information Screen –
Inventory Data References Table
Portion**

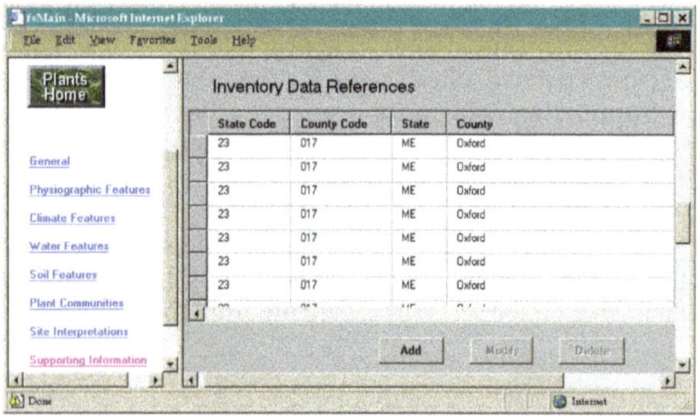

The part of the **Supporting Information** screen shown in Figure 637-81 displays the Inventory Data References table that lists the site inventory plots supporting the site description. It contains the following items:

Add/Modify – Click on the Add or Modify button at the bottom of the table to record new or modify existing data. For new entries, the screen shown in Figure 637-82 will be displayed. When modifying previously entered data, a screen identical to the screen shown in Figure 637-82 will be displayed, except the screen is titled Modify rather than Add.

Enter a descriptive identifier for the source of the plot data in the Data Source field. For example, site data recorded in the NRCS forest site inventory database should be referenced as ESI–Forestland. In the Number, Year, State, and County fields, enter the data that corresponds to the plot ID recorded in the ESI–Forestland database. The state and county are selected from the choice list provided.

After entering the appropriate data, click on the Save button to save the data to the database, then continue entering other data references, if appropriate. When all data have been entered, click on the Return button to return to the **Supporting Information** screen. Click on the Clear button to clear all entries. The Clear button will not delete previously saved entries.

Figure 637-82   Forest Data References Screen

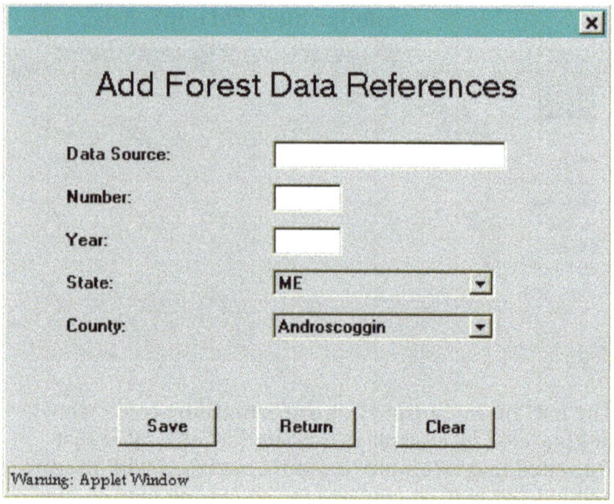

Figure 637-83   Supporting Information Screen – State Correlation Portion

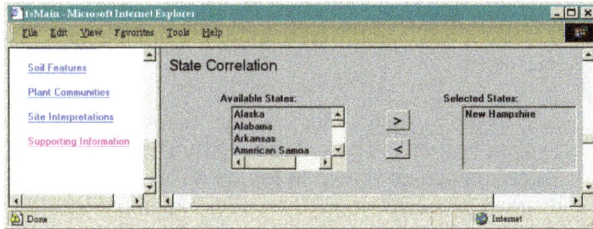

The part of the **Supporting Information** screen shown in Figure 637-83 is used to list other states using the site description of the site being described. It contains the following items:

Available States – For each state that uses the site being described, click on the state name in the choice list and then click on the > button to add that state to the selected states list. To remove a state from the selected states list, click on the < button.

Figure 637-84   Supporting Information Screen – Type Locality Portion

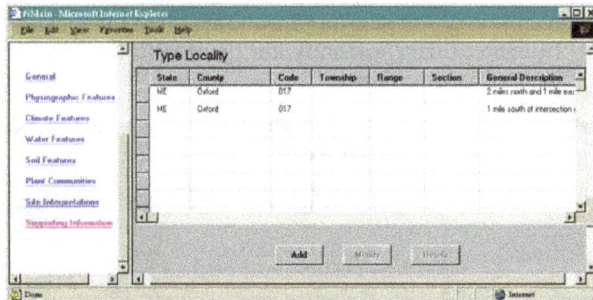

The part of the **Supporting Information** screen part shown in see Figure 637-84 displays the Type Locality table that lists the physical location of sites that typify the site being described. It contains the following items:

Add/Modify – Click on the Add or Modify button at the bottom of the table to record new or modify existing data. For new entries, the screen shown in Figure 637-85 will be displayed. When modifying previously entered data, a screen identical to the screen shown in Figure 637-85 will be displayed, except the screen is titled Modify rather than Add.

Figure 637-85   Type Localities Screen

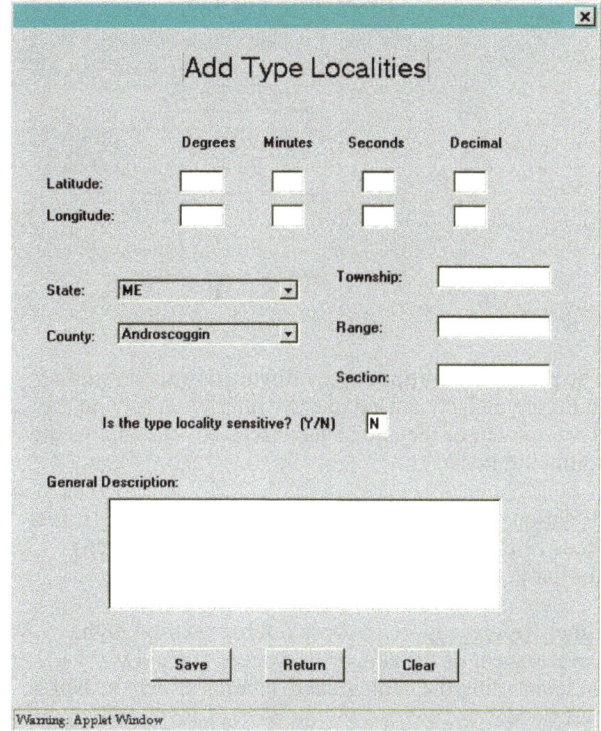

Enter the latitude and longitude of each typifying site. The township, range, section, and a general description of the location may also be recorded. If entered, the state and county are selected from the choice list provided.

After entering the appropriate data, click on the Save button to save the data to the database, then continue entering type localities, if appropriate. When all data have been entered, click on the Return button to return to the **Supporting Information** screen. Click on the Clear button to clear all entries. The Clear button will not delete previously saved entries.

Delete – Highlight a row in the Type Locality table and click on the Delete button at the bottom of the table to delete all data for that particular row from the database.

637-39

**Figure 637-86   Supporting Information Screen –
Relationship Portion**

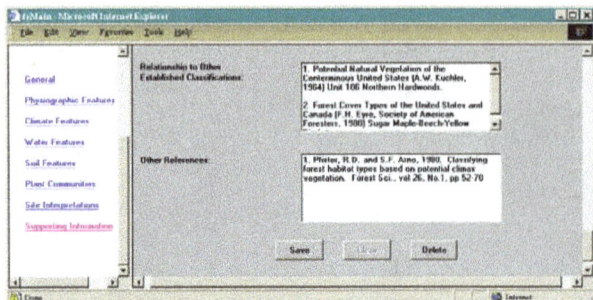

The part of the **Supporting Information** screen shown in Figure 637-86 is used to enter references used in the development of the site being described. It contains the following items.

Relationship to Other Established Classifications – List those references that describe classification systems similar to the site being described.

Other References – List those references used in the development of the site, or references that aid in understanding the ecological dynamics of the site being described.

(3) Producing Ecological Site Description Reports
 Ecological Site Descriptions stored in the database are available for viewing through pre-formatted reports. These reports are formatted to resemble the sample report shown in the National Forestry Manual, exhibit 537-16. The user has the option of producing a complete Ecological Site Description report or an individual report of any one of the eight sections described above. Ecological Site Description reports are available to anyone with access to the Internet. A password or user name is not required to produce ESD reports.

To produce an ESD report, complete the following on the ESD homepage (see Figure 637-40):

1. Select a State – From the choice list, select the appropriate state. You must select a state to produce an ESD report.

2. Select an MLRA – You may choose to select an appropriate MLRA from the choice list, but it is not required. Selecting an MLRA serves to reduce the

number of ESD's from which to choose, if the MLRA covers only a portion of the state selected.

3. Select a Function – Click on the Reports radio button. You do not need to enter a username or password.

4. Access the System – Click on the Access the ESD System button. The **Ecological Site Description Selection** screen shown in Figure 637-45 will be displayed. Select an ESD by clicking on the ID number. A screen similar to the one shown in Figure 637-87 will be displayed.

5. Select Report – The **Report** screen (see Figure 637-87) lists the report options in the left frame of the screen. Upon initial display of the **Report** screen, the General section of the report is displayed in the right-hand frame. To view other sections of the report, click on the section name in the left-hand frame. To view the entire ESD report, click on the Complete Report option.

6. Print Report – To print the report displayed in the right frame, use the browser's print function.

NOTE: Due to current limitations in the technology for printing html documents from the Internet, the report may not print with the desired pagination.

Figure 637-87   ESD Report Screen

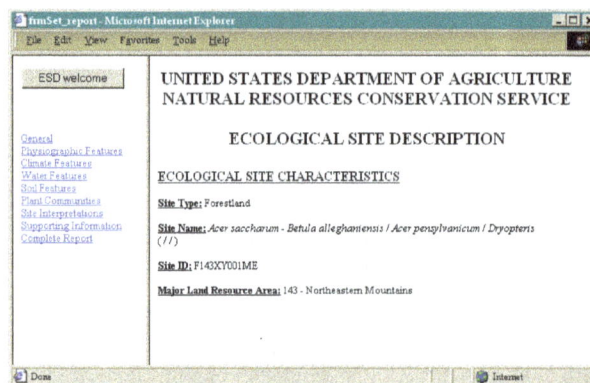

637.32 Ecological Site Inventory

An inventory is defined as the collection, assemblage, interpretation, and analysis of natural resource data. The data are used to monitor ecological change, develop ecological site descriptions, determine soil productivity and species adaptability, and predict height growth and species survival rates.

In the Natural Resources Conservation Service (NRCS), ecological site inventories are always correlated to a specific soil component. In NRCS, the work of correlating site inventories to the soil began in the middle 1940's and thousands of site inventories have been made since that time. Early inventories were primarily concerned with relating soils to forest productivity (site index). NRCS now recognizes the need for a more comprehensive inventory of the resources in order to document the relationship between the soil and plant composition, in both the overstory and understory.

The National Forestry Manual, Part 537 defines policies and standards for collection of the data associated with forest ecological site inventories. This subpart provides detailed instructions on the collection of forest plot data and the use of the Ecological Site Inventory-Forestland application to enter, edit, and report out the data associated with forest inventories.

(a) Forest Plot Inventory Operations

The National Forest Manual, Part 537 details the standards to be followed in the collection of ecological site inventory data. This subpart details the procedures used in the collection of forest plot inventory data.

(1) Site Selection
The quality of the inventory is dependent upon the quality of the site where the data are collected. Plot selection involves team work, usually consisting of a forester and a soil scientist. Both members of the team must be satisfied that the plot is satisfactory before any data are collected. Consider the following when selecting sites.

- Stand to be inventoried need not cover a large area.

- Trees should be in good health and have the appearance of having grown under normal conditions.

- Stands that appear stagnated should not be sampled, nor should open-grown, scattered trees.

- The trees measured should be growing under soil conditions that appear the same for all trees measured in a particular plot.

- A tree should not be used if the examination of the increment core reveals one or more periods of suppression, or if the tree shows evidence of ice breakage or other damage to the terminal during the life of the tree.

- Stands less than 30 years of age do not ordinarily give accurate site index values

- The soil must be modal for the taxonomic unit.

- All trees to be measured must be on the selected soil and on the same soil conditions – within ± 3% of the same slope and within ± 15 degrees of the same aspect.

(2) Site Documentation
The soil scientist describes the soil profile using standard terms and abbreviations as detailed in Chapters 3 and 5 in Agriculture Handbook 18, Soil Survey Manual, October 1993. Attach a copy of the profile description to the ESI Forest Plot Field Worksheet.

The forester records field data on the ESI Forest Plot Field Worksheet. These worksheets should be filed for safekeeping and future reference.

All ESI–Forestland field data are to be entered into the ESI–Forestland database. Instructions and procedures for entering the field data using the ESI–Forestland application are contained in this subpart.

(3) ESI Forest Plot Field Worksheet
The ESI Forest Plot Field Worksheet is used to record field data. The worksheet is designed to closely match the data entry screens in the ESI–Forestland application. Exhibit 637-55 shows an example of a completed worksheet.

A blank copy of the ESI Forest Plot Field Worksheet (including instructions and codes) that can be printed and reproduced for use in the field is available on the NFH Web site at http://soils.usda.gov/technical/nfhandbook. To obtain these documents, click on Exhibit 637-55, Exhibit 637-56, and Exhibit 637-57 in Part 637 of the NFH Web site. Use the print function of your Internet browser to print the documents. The documents can also be saved to your local computer for subsequent printing.

The following are detailed instructions for recording plot data on the ESI Forest Plot Field Worksheet.

(i) *Forest Site Plot Number*
The plot number has four parts:

ID – Enter a number from 1 to 999. It is suggested that numbers be used in chronological order, by year, in each county.

Year – Enter the last two digits of the year.

State and County – Enter the FIPS 2-digit numerical code for the state and the FIPS 3-digit code for the county.

(ii) *Location Data*
Location Description – Describe the location by entering the approximate distance the plot is from the closest section corner: (e.g., the plot is 650 feet west and 1,500 feet north of the southeast corner of Section 12, T25N, Rl2W) or by entering the distance and direction from a reasonably permanent local landmark, if the area is not sectionalized).

Cover Type – Enter the appropriate forest cover type code. See Exhibit 637-59 for a list of forest cover type codes. A full description of each forest cover type and its components is found in the source manual: *Forest Cover Types of the United States and Canada* (F.H. Eyre, Society of American Foresters. 1980).

MLRA Number – Enter the major land resource area number according to the map and descriptions in Agriculture Handbook 296.

State Plane Coordinates –

- *Zone* – In Alaska (zones 1-10), California (zones 1-7), and Hawaii (zones 1-5) zones are numbered and the number is the code for the zone. The 10th zone of Alaska is coded 0. In Connecticut, Delaware, Maryland, New Hampshire, New Jersey, North Carolina, Rhode Island, Tennessee, and Vermont, the entire state is one zone: code 0. For all other zone codes, refer to Exhibit 637 - 58.

- *East and North* – Enter the East (x) and North (y) coordinates. Many county highway maps and later U.S. Geological Survey (USGS) quad sheets have

state plane coordinate tick marks along the edges. Use USGS quad sheets whenever possible. When tick marks are present, the coordinates can be determined by use of an appropriate scale. Transparent cross-sectional scales for the 7-1/2 minute and 15-minute quad sheets may be ordered from commercial forestry supply catalogs. If the county highway map has lines of latitude and longitude, these may be used to determine coordinates. The U.S. Department of Commerce Coast and Geodetic Survey has published "Plane Coordinate Intersection Tables" for each state. These publications are available from the Government Printing Office at a nominal cost. For each 2-1/2 minute intersection of longitude and latitude, the x and y coordinates are given. An intersection may be used as a base point from which coordinates can be measured using a suitable scale. Coordinates should be rounded to show the probable accuracy.

Section/Range/Township –

- *Section* – Enter the number of the section in which the plot is located.

- *Range* – Enter the range designation in the manner: R4E or R4W. Use no other abbreviations.

- *Township* – Enter the township designation in the manner: T7N or T6S. Use no other abbreviations.

Elevation – Enter the elevation in feet. Elevation may be read from a contour map, obtained by use of a barometer, or estimated from known reference points. Where there is little variation is elevation, as in the coastal plain and piedmont sections of the southeast, only the general elevation of the area is needed. Do not take extra time to obtain accuracy closer than 100 feet in mountainous country.

(iii) *Physical Data*
Precipitation – Enter the annual precipitation and summer (warm season) precipitation, each to the nearest whole inch. The VegSpec application provides a convenient source for obtaining precipitation data. The VegSpec database contains climate data provided by the NRCS National Water and Climate Center in Portland, Oregon for hundreds of climate stations. Access the VegSpec application by clicking on the VegSpec link on the Plants homepage at http://plants.usda.gov. To view the climate data, start the VegSpec application, select a state and a climate station on the Site Description screen,

and then click on the "climate attributes sub-option" hyperlink.

<u>Landform</u> – Enter one of the following codes.

Landform	Code	Landform	Code
Swamp	SM	Level and Undulating Plains/Plateaus	LP
Basins, Playas, Lakebeds	B	Rolling and Hilly Plains/Plateaus	RP
Flood Plains, Bottoms	FP	Mountains, Steep Hills, Dissected Plateaus	M
Stream Terraces	ST	Sand Dunes, Sand Hills	SD
Fans, Alluvial, Colluvial	F	Flatwoods	FL

<u>Slope</u> –

- *Percent* – Enter percent of slope in whole numbers, 0 to 99. Use 99 for slopes steeper than 99.

- *Kind* – Enter one of the following codes.

Slope Kind	Code
Plane, Single, or Simple	P
Irregular or Complex	I

- *Shape* – Leave blank if kind of slope is coded "P." Use one of the following codes if kind of slope is coded "I"

Slope Shape	Code	Slope Shape	Code
Concave Horizontally	1	Convex Vertically	4
Concave Vertically	2	Concave	5
Convex Horizontally	3	Convex	6

- *Microrelief* – Enter one of the following codes, or leave blank if none apply.

Microrelief	Code	Microrelief	Code
Cradle Knoll	C	Tree Mounds	T
Gilgai	G	Mounded	M
Frost Polygons	F		

- *Aspect* – Enter the direction the slope faces in degrees of azimuth, clockwise from true north.

- *Length* – Enter the approximate length of slope in even feet. Round to indicate probable accuracy.

- *Position on Slope* – Enter one of the following codes.

Slope Position	Code	Slope Position	Code
Lower Slope	L	Middle Slope	M
Upper Slope	U	Noninfluencing	X

(iv) **Soil Data**

<u>Detailed Profile</u>– Enter Y if a complete and suitable soil description has been made. If not, enter N. The value of the work may be discredited if there is not a complete and suitable description meeting modern standards.

<u>Detailed Understory</u> – If detailed understory information has been recorded for some related purpose, enter Y. If not, enter N.

<u>Mensurational Information</u> – If detailed mensurational information has been recorded for the trees in the plot, enter Y. If not, enter N. Some soil-site work is conducted in cooperation with other agencies or companies that carry on complete mensurational studies. If this is the case, reference should be recorded in Remarks, e.g., U. of N.C. plot 7-w.

<u>Soil Series Name</u> – Enter the full series name. If a series name has not been established for the soil, enter the series name of the family to which the soil belongs, followed by the letter (F). Example: Holden (F).

<u>Texture</u> –

- *Modifier* – Enter one of the following codes, or leave blank if none apply.

Texture Modifier	Code	Texture Modifier	Code
Ashy	ASHY	Herbaceous	HB
Bouldery	BY	Hydrous	HYDR
Boulderymucky	BYMK	Medial	MEDL
Very bouldery	BYV	Mucky	MK
Very boulderymucky	BYVMK	Marly	MR
Extremely bouldery	BYX	Mossy	MS
Extremely boulderymucky	BYXMK	Parabouldery	PBY
Cobbly	CB	Very parabouldery	PBYV
Cobblymucky	CBMK	Extremely parabouldery	PBYX
Very cobbly	CBV	Paracobbly	PCB
Very cobblymucky	CBVMK	Very paracobbly	PCBV
Extremely cobbly	CBX	Extremely paracobbly	PCBX
Extremely cobblymucky	CBXMK	Parachannery	PCN
Channery	CN	Very parachannery	PCNV
Very channery	CNV	Extremely parachannery	PCNX
Extremely channery	CNX	Permanently frozen	PF
Coprogenous	COP	Paraflaggy	PFL
Diatomaceous	DIA	Very paraflaggy	PFLV
Flaggy	FL	Extremely paraflaggy	PFLX
Very flaggy	FLV	Paragravelly	PGR
Extremely flaggy	FLX	Very paragravelly	PGRV
Gravelly	GR	Extremely paragravelly	PGRX
Coarse gravelly	GRC	Parastony	PST
Fine gravelly	GRF	Very parastony	PSTV
Coarse gravellymucky	GRCMK	Extremely parastony	PSTX
Fine gravellymucky	GRFMK	Peaty	PT
Medium gravelly	GRM	Stratified	SR
Gravellymucky	GRMK	Stony	ST
Very gravelly	GRV	Stonymucky	STMK
Very gravellymucky	GRVMK	Very stony	STV
Extremely gravelly	GRX	Very stonymucky	STVMK
Extremely gravellymucky	GRXMK	Extremely stony	STX
Grassy	GS	Extremely stonymucky	STXMK

Texture Modifier	Code	Texture Modifier	Code
Gypsiferous	GYP	Water	W

Wetted	W
Drained	D

- *Type* – Enter one of the following codes, or leave blank if none apply.

Texture Type	Code	Texture Type	Code
Clay	C	Sand	S
Clay loam	CL	Sandy clay	SC
Coarse sand	COS	Sandy clay loam	SCL
Coarse sandy loam	COSL	Silt	SI
Fine sand	FS	Silty clay	SIC
Fine sandy loam	FSL	Silty clay loam	SICL
Loam	L	Silt loam	SIL
Loamy coarse sand	LCOS	Sandy loam	SL
Loamy fine sand	LFS	Very fine sand	VFS
Loamy sand	LS	Very fine sandy loam	VFSL
Loamy very fine sand	LVFS		

- *Terms In Lieu of Texture* – Enter one of the following codes, or leave blank if none apply.

Terms In Lieu of Texture	Code	Terms In Lieu of Texture	Code
Bedrock	BR	Petrocalcic	PC
Boulders	BY	Paracobbles	PCB
Cobbles	CB	Parachanners	PCN
Channers	CN	Peat	PEAT
Duripan	DUR	Petroferric	PF
Flagstones	FL	Paraflagstones	PFL
Gravel	G	Paragravel	PG
Highly decomposed plant material	HPM	Petrogypsic	PGP
Material	MAT	Placic	PL
Moderately decomposed plant material	MPM	Parastones	PST
Mucky peat	MPT	Slightly decomposed plant material	SPM
Muck	MUCK	Stones	ST
Ortstein	OR	Water	W
Paraboulders	PBY		

Past Erosion – Enter one of the following codes, or leave blank if none apply.

Past Erosion Class	Code
Moderate (Classes 1 and 2)	1
Severe (Classes 3 and over)	2

Drainage Class – Enter one of the following codes, or leave blank if none apply.

Drainage Class	Code	Drainage Class	Code
Very Poorly Drained	1	Well Drained	5
Poorly Drained	2	Somewhat Excessively Drained	6
Somewhat Poorly Drained	3	Excessively Drained	7
Moderately Well Drained	4		

Altered Water Relations – Enter one of the following codes, or leave blank if there has been no change in the water regime.

Altered Condition	Code

(v) Density Data

Understory Abundance – Rate each item (Reproduction, All Woody Plants, Grasses and Forbs, Mosses and Lichens) according to their relative abundance. Enter one of the following codes.

Abundance	Code
None	1
Sparse	2
Moderately Abundant	3
Abundant	4
Very Abundant	5

Stand Density (Canopy) –

- *M. or E.* – Enter a code that indicates if the crown canopy closure was measured (M) or estimated (E).

- *Percent* – Enter the percent of crown canopy closure in whole numbers.

Basal Area –

- *M. or E.* – Enter the code that indicates if the basal area was measured (M) or estimated (E).

- *Sq. Ft.* – Enter basal area in square feet.

Crown Competition Factor – Enter the CFC number. This data element applies only to lodgepole pine. See Exhibit 637-59 for instructions on determining site index for lodgepole pine.

(vi) Tree Data

Ordinarily five to six trees of a single species are measured to determine site index. The information for one tree is placed on one line. If more than one species is measured, it is helpful to group the trees by species. If more trees are measured than the space will accommodate, use a second worksheet. Fill in the plot number for identification and continue measurements on the second worksheet.

NSPNS – Enter the scientific plant name symbol as listed in the Plants database (http://plants.usda.gov).

Crown Class – Enter the following crown class code for each measured tree.

Crown Class	Code
Dominant	D
Codominant	C

Tree Origin – Enter the following tree origin code of each measured tree.

Tree Origin	Code
Naturally Seeded	S
Planted	P
Coppice	C
Direct Seeded	D

Tree Diameter – Enter the diameter at breast height of each measured tree to the nearest 0.1 inch.

Inches Radius Last 10 Years – Enter the radius of the last 10 rings of each measured tree to the nearest 0.1 inch.

Age Estimation –

- *Ht. Ring Count.* – Enter the height, in feet, of each measured tree, at which the growth rings were counted. For standing trees this is normally at 4.5 feet above normal ground level. If ring count is from stumps, the center pith of the tree should be clearly discernable in the core sample to insure an accurate age count.

- *No. of Rings* – Enter the number of growth rings counted. When an increment borer is being used to determine age, certain precautions are important. Most trees are elliptical, with the pith displaced from the center along the long axis and away from the direction of lean. Size the tree up carefully, aim the borer with care, and bore in the direction of the long axis to ensure reaching the pith. If the borer misses the pith, a second boring is preferable to estimating the number of rings to pith. Care in selecting the point to bore can eliminate most errors.

- *Mea. Pt. Age* – Enter the number of years it took for the measured tree to reach the height at which the ring count is made. To determine age with an increment borer, bore all trees at breast height and add standard age correction factors. This reduces some of the errors caused by variations in growth rate during the regeneration period. An approximate correction factor is added to the increment core age and this total together with tree height is used to determine approximate site index. Using the approximate site index, the true age correction factor is obtained. This factor added to increment core age gives true total age. Use only the age correction factors contained in the National Forestry Manual, Exhibit 537-1. Do not use correction factors for

sprouts. If only a breast height age curve is used to determine site index for a species, leave this entry blank.

- *Total Age* – Enter the total age. The total age is the sum of the number of rings plus the age correction factor. When site index is determined by using a breast height age curve, make no entry in the total age data field. If both total age and breast height age curves are used for a species, enter total age.

- *Total Height* – Enter the total height of the tree to the nearest foot. Care is essential to obtain accurate height measurements. Any of several instruments may be used (Abney level, Haga altimeter, clinometer, etc). Refer to Part 636.3 for information on measuring instruments. It is good practice to locate a point from which the top and base of the tree can be seen clearly. Take readings from this point, then measure back to the tree to get the distance. This procedure is more accurate and less time consuming than measuring a set distance out from the tree often to find that the point does not provide an unobstructed view. The measuring point should be at least as far from the tree as the tree is tall.

(vii) Site Index Data
This section is used to record the site index for the species measured.

NSPNS – Enter the scientific plant name symbol as listed in the Plants database (http://plants.usda.gov).

Number of Trees – Enter the number of trees used to calculate the site index.

Site Index Curve Number – Select the code number for the site index curve used to derive the site index. For a list of site index curve numbers, refer to the National Forestry Manual, Exhibit 537-1.

Average Site Index – Enter the site index for each species measured. From the Tree Data Section, calculate average total age and average total height. Use the appropriate site index curves in the standard manner to obtain site index. Since the age correction factor depends on the site index, a two-step process is necessary: (1) average the number of rings and add an estimated age with which to calculate an approximate site index, then (2) use the true age correction factor to obtain total age and true site index.

637-45

(viii) Canopy Cover Data

NSPNS – Enter the scientific plant name symbol as listed in the Plants database (http://plants.usda.gov). If there are more than twelve species, combine the least important species using the code "OTHER".

Percent – Enter the percent, in whole numbers, that each species occupies of the total canopy. The percentages must total 100. Use the code "OTHER" (see "NSPNS", above) if more than 12 species occur in the canopy, then record the percentage needed to sum to 100. Refer to Part 636.3 for information on various instruments used to measure canopy cover.

(ix) Ground Cover Data

NSPNS – Enter the scientific plant name symbol as listed in the Plants database (http://plants.usda.gov) for the species that make up the ground cover below the canopy (mostly under 15 feet).

Rating – Enter the following code that represents the relative abundance of each listed species.

Abundance	Rating Code
An Occasional Plant Present	1
Sparse	2
Moderately Abundant	3
Abundant	4
Very Abundant (at least 30 percent of ground cover)	5

(xi) Remarks

Enter any pertinent information about the site.

(4) Statistical Analysis

Statistical analysis of forest plot site index data is useful in determining the reliability of the estimate of site index. Common measures of reliability for site index estimates are range, mean, standard deviation, and coefficient of variation. For plot data entered into the ESI–Forestland database, the range, mean, and standard deviation are available through the standard report option of the ESI–Forestland application.

When plot data have not been entered into the ESI–Forestland database or it is desired to determine standard deviation for finer divisions of populations than is possible through the ESI–Forestland standard report option, the following procedures may be used.

(i) Standard Deviation

To determine the standard deviation of site index, group the forest plot site index data by soil series and tree species. If desired, the data may be further divided by state, county, MLRA, precipitation zones, latitudinal belts, etc. If the forest plot data have been entered in the ESI–Forestland database, the custom report option of the ESI–Forestland application is a convenient method to obtain site index data for various divisions.

To calculate standard deviation manually, use the formula:

$$\text{Standard Deviation} = \sqrt{\frac{\sum I^2 - \frac{(\sum I)^2}{n}}{(n-1)}}$$

Where:
I = site index for a plot
n = number of plots

The following table illustrates the use of the above formula to calculate site index for a population of 15 forest site plots in two counties.

Number	County	Year	Plant Symbol	Site Index (I)	Site Index Squared (I^2)
1	67	55	PIEC2	77	5929
4	67	55	PIEC2	60	3600
5	67	55	PIEC2	66	4356
6	67	55	PIEC2	61	3721
1	37	76	PIEC2	73	5329
18	37	75	PIEC2	69	4761
20	37	75	PIEC2	62	3844
20	37	76	PIEC2	68	4624
21	37	75	PIEC2	73	5329
21	37	76	PIEC2	70	4900
23	37	76	PIEC2	75	5625
26	37	75	PIEC2	74	5476
29	37	75	PIEC2	72	5184
30	37	76	PIEC2	69	4761
31	37	76	PIEC2	64	4096
n=15				1033	71535

$$\text{Standard Deviation} = \sqrt{\frac{71535 - \frac{1033 \times 1033}{15}}{14}} = \pm 5.3$$

Another convenient and simple method of calculating standard deviation is to enter the forest plot site data into an electronic spreadsheet, such as Microsoft® Excel, then use the built-in statistical function to calculate standard deviation. If the plot data have been generated through the custom report option of the ESI–Forestland application, this data may be pasted directly into an electronic spreadsheet, eliminating the need to manually enter the data. Once the data are pasted into the worksheet, the standard deviation can be easily calculated for the whole population or any desired subset. Figure 637-88 shows an example of calculating standard deviation using an electronic spreadsheet. Notice that the built-in function, "STDEV(E2:E16)", is used to calculate the standard deviation.

soil component. Since that is not normally possible, site index data taken on the same soil component in different work areas are used to derive site index. Each block of plots needs to be scrutinized carefully to see if the plots represent reasonably homogeneous conditions of elevation, precipitation, aspect, and latitude.

It might be reasoned that the plot data shown in Figure 637-88 are quite homogeneous since the standard deviation is well below 10 percent. Closer scrutiny, however, shows that most of the plots from County 67 have site indexes lower than the mean for the combined counties, whereas those from County 37 have site indexes above the mean for the combined counties. An analysis of each county is shown in the following tables.

Figure 637-88   Calculating Standard Deviation using Electronic Spreadsheet

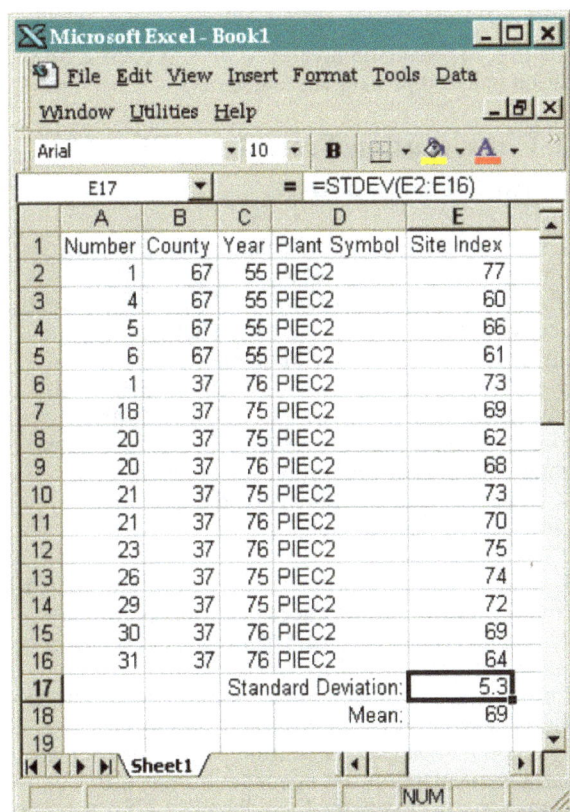

Number	County	Year	Plant Symbol	Site Index
1	67	55	PIEC2	77
4	67	55	PIEC2	60
5	67	55	PIEC2	66
6	67	55	PIEC2	61
			Standard Deviation	7.8
			Mean	66

Number	County	Year	Plant Symbol	Site Index
1	37	76	PIEC2	73
18	37	75	PIEC2	69
20	37	75	PIEC2	62
20	37	76	PIEC2	68
21	37	75	PIEC2	73
21	37	76	PIEC2	70
23	37	76	PIEC2	75
26	37	75	PIEC2	74
29	37	75	PIEC2	72
30	37	76	PIEC2	69
31	37	76	PIEC2	64
			Standard Deviation	4.1
			Mean	70

Separating the plots on a county basis seems to be desirable in this case, because the standard deviation is lowered by doing so. Climatic differences may occur. The person doing the work should be satisfied that the separation is a reasonable one and not the product of some other unmentioned factor.

The standard deviation can be used to check the validity of any plot in a population. Any plot with a site index more than two and one-half standard deviations from the mean of the population is likely not in the same universe

It would be ideal if sufficient numbers of plots could be taken in every working area to establish the relationship of site index for each of the various tree species to each

or population and can justifiably be excluded from the population.

Where few data exist, plots with fewer than the minimum number of trees (5-6) may need to be included in the population for determination of site index. Although the site index and standard deviation may be determined in the normal manner, it may be desirable to determine the site index and standard deviation by weighting the plots according to the number of trees measured for each plot, rather than the number of plots. The following is an example of the procedure. Note that n is now the number of trees rather than the number of plots.

Number	County	Year	Plant Symbol	Site Index (I)	Trees Measured (n)	$I \times n$
1	67	55	PIEC2	77	1	385
4	67	55	PIEC2	60	5	300
5	67	55	PIEC2	66	5	330
6	67	55	PIEC2	61	3	183
1	37	76	PIEC2	73	1	73
18	37	75	PIEC2	69	2	138
			Totals	406	17	1101

Average site index on plot basis: $\dfrac{406}{6} = 68$

Average site index on individual tree basis: $\dfrac{1101}{21} = 65$

In calculating the standard deviation, the weighting principle is continued as demonstrated in the following example.

Site Index (I)	Trees Measured (n)	$I \times n$	$(I \times n)^2$
77	1	385	5929
60	5	300	18000
66	5	330	21780
61	3	183	11163
73	1	73	5329
69	2	138	9522
406	17	1101	71723

$$\text{Standard Deviation} = \sqrt{\frac{\sum Ixn^2 - \dfrac{(\sum Ixn)^2}{n}}{(n-1)}}$$

$$\text{Standard Deviation} = \sqrt{\frac{71723 - \dfrac{1101x1101}{17}}{16}} = \pm 5.1$$

(ii) Coefficient of Variation

Another statistical measure commonly used is coefficient of variation. Coefficient of variation is the standard deviation expressed as a percentage of the mean:

$$\text{Coefficient of Variation} = \left(\frac{\text{Standard Deviation}}{\text{Mean}}\right) x100$$

Ordinarily, in forestland inventory work, a coefficient of variation below 8 percent is satisfactory.

(b) Conservation Tree/Shrub Plot Inventory Operations

NRCS documentation of conservation tree/shrub species performance began in the 1950's. Procedures were developed and improved during the 1960's when a great number of evaluation plots were established in the Great Plains States. The objective of the inventories is to study and record the relations between the kinds of soil and the performance of individual tree and shrub species. The information is used in selecting species best suited for different kinds of soil; predicting longevity, effectiveness, and height growth; and improving spacing, design, and renovation.

Soil, climate, and physiographic site factors are studied and recorded by commonly accepted techniques. Determination of the influence of site factors is one of the most important and difficult aspects of the study. Performance is determined insofar as possible under the "standard" conditions of management and species competition that have resulted from particular spacings and arrangements. It is not intended to study the conservation effectiveness of windbreaks. With accurate information about individual species performance under specified conditions of soil, climate, physiography, and management, including species, spacing, and arrangement, the relative potential effectiveness of different kinds of windbreaks can be inferred.

637-48

(1) Site Selection

Sites are tentatively selected for study on the basis of age, soil on which they are growing, management, and prospects of providing valid information on species performance. The plantings may be of any age and should show evidence of reasonably good management. Species should be measured at 5-year intervals until they reach their average mature height.

Plot selection involves teamwork, usually consisting of a plant scientist and a soil scientist. Both members of the team must be satisfied that the plot is satisfactory before any data are collected.

A soil scientist makes certain that each sampled area (plot) is representative of a soil taxonomic unit. Each soil taxonomic unit sampled is to be within the range of characteristics of each described and named soil studied. Concentrating studies on "benchmark" soils, where this is possible, is one way the soil scientist can project known information to similar soils.

A plant scientist makes certain that each row segment selected for study provides accurate and useful information on individual species performance under the conditions to be recorded.

Plots need not be permanently marked, but the description should permit the precise identification of each plot. Periodic measurement of some plantings provides valuable information on the development of specific trees under identical circumstances over a period of time.

For windbreaks, a plot consists of a series of segments of the rows of individual species on a single kind of soil. The row segments of the different species are usually adjacent across the planting. For each row segment studied, conditions within the row and in adjacent rows must be such that performance to be recorded will accurately reflect conditions of competition, species, spacing, and arrangement.

A given row segment may not provide valid information when plants are missing in the row or in adjacent rows because competition is reduced within the row or between rows; thus, these segments are not indicative of species performance. If trees are missing and suspected never to have been established, some offsetting of row segments to be measured may be needed. The offsetting is to be limited to a minor portion of the planting, and the different row segments are to be approximately adjacent within the windbreak.

Confine observations and measurements on each row to a row segment of 10 to 25 original plant spaces.

(2) Site Documentation

The soil scientist describes the soil profile using standard terms and abbreviations as detailed in Chapters 3 and 5 in Agriculture Handbook 18, Soil Survey Manual, October 1993. Attach a copy of the profile description to the ESI Windbreak Plot Field Worksheet.

The plant scientist records field data on the ESI Windbreak Plot Field Worksheet. These worksheets should be filed for safekeeping and future reference.

All plot field data are to be entered into the ESI–Forestland database. Instructions and procedures for entering the field data using the ESI–Forestland application are contained in this subpart.

(3) ESI Windbreak Plot Field Worksheet

The ESI Windbreak Plot Field Worksheet is used to record field data. The worksheet is designed to closely match the data entry screens in the ESI–Forestland application. Exhibit 637 - 58 shows an example of a completed worksheet.

A blank copy of the ESI Windbreak Plot Field Worksheet (including instructions and codes) that can be printed and reproduced for use in the field is available on the NFH Web site at http://soils.usda.gov/technical/nfhandbook. To obtain these documents, click on Exhibit 637-60, Exhibit 637-61, and Exhibit 637-62 in Part 637 of the NFH Web site. Use the print function of your Internet browser to print the documents. The documents can also be saved to your local computer for subsequent printing.

The following are detailed instructions for recording plot data on the ESI Windbreak Plot Field Worksheet.

(i) General Data

The plot number has four parts: ID, year, state and county.

ID – Enter a number from 1 to 999. It is suggested that numbers be used in chronological order, by year, in each county.

Year – Enter the last two digits of the year.

State and County – Enter the FIPS 2-digit numerical code for the state and the FIPS 3-digit code for the county.

637-49

East	E
West	W
Northwest	NW
Southwest	SW
Southeast	SE
Northeast	NE

P-E Index – Enter the P-E index (precipitation-evaporation index). Great Plains states use drawing number 5S-16, 270 to determine index. Other states can leave this field blank.

Precipitation – Enter the total annual precipitation and summer (warm season) precipitation, each to the nearest whole inch. The VegSpec application provides a convenient source for obtaining precipitation data. The VegSpec database contains climate data provided by the NRCS National Water and Climate Center in Portland, Oregon for hundreds of climate stations. Access the VegSpec application by clicking on the VegSpec link on the Plants homepage at http://plants.usda.gov/. To view the climate data, start the VegSpec application, select a state and a climate station on the Site Description screen, and then click on the climate attributes sub-option hyperlink.

Extra Moisture – Enter the following applicable moisture source code.

Moisture Source	Code
None	N
Irrigation	I
Water Table	W
Contour Planting	C
Position	P

In Vital Notes, record the following relative to the moisture source code entered:
Code I - how long the planting was irrigated, if irrigation has been provided for less than the life of the planting.
Code W - indicate depth in inches to permanent water table.
Code C or P - indicate foot slope, drain bank, etc.

Elevation – Enter the elevation in feet. Elevation may be read from a contour map, obtained by use of a barometer, or estimated from known reference points. Where there is little variation in elevation, as in the Coastal Plain and Piedmont sections of the Southeast, only the general elevation of the area is needed. Do not take extra time to obtain accuracy closer than 100 feet in mountainous country.

MLRA Number – Enter the major land resource area number according to the map and descriptions in Agriculture Handbook 296.

Direction Row 1 Faces – Enter the following applicable code.

Row 1 Direction	Code
North	N
South	S

State Plane Coordinates –

* Zone – In Alaska (zones 1–10), California (zones 1–7), and Hawaii (zones 1–5) zones are numbered and the number is the code for the zone. The 10th zone of Alaska is coded 0. In Connecticut, Delaware, Maryland, New Hampshire, New Jersey, North Carolina, Rhode Island, Tennessee, and Vermont, the entire state is one zone: code 0. For all other zone codes, refer to Exhibit 637 - 58.

* East and North – Enter the East (x) and North (y) coordinates. Many county highway maps and later U.S. Geological Survey (USGS) quad sheets have state plane coordinate tick marks along the edges. Use USGS quad sheets whenever possible. When tick marks are present, the coordinates can be determined by use of an appropriate scale. Transparent cross-sectional scales for the 7-1/2 minute and 15-minute quad sheets may be ordered from commercial forestry supply catalogs. If the county highway map has lines of latitude and longitude, these may be used to determine coordinates. The U.S. Department of Commerce's Coast and Geodetic Survey has published "Plane Coordinate Intersection Tables" for each state. These publications are available from the Government Printing Office at a nominal cost. For each 2-1/2 minute intersection of longitude and latitude, the x and y coordinates are given. An intersection may be used as a base point from which coordinates can be measured using a suitable scale. Coordinates should be rounded to show the probable accuracy.

Section/Range/Township –

* Section – Enter the number of the section in which the plot is located.

* Range – Enter the range designation in the manner: R4E or R4W. Use no other abbreviations.

* Township – Enter the township designation in the manner: T7N or T6S. Use no other abbreviations.

(ii) Soil Information

Series Name – Enter the full soil series name. If a soil series name has not been established for the soil, enter the series name of the family to which the soil belongs, followed by the letter (F). Example: Holden (F).

Type – Select the following appropriate texture type code, or leave blank if none apply.

Texture Type	Code	Texture Type	Code
Sand	S	Silt	SI
Coarse Sand	COS	Sandy Clay Loam	SCL
Fine Sand	FS	Clay Loam	CL
Very Fine Sand	VFS	Silty Clay Loam	SICL
Loamy Coarse Sand	LCOS	Sandy Clay	SC
Loamy Sand	LS	Clay	C
Loamy Fine Sand	LFS	Silty Clay	SIC
Loamy Very Fine Sand	LVFS	Fibric Materials	FB
Coarse Sandy Loam	COSL	Hemic Materials	HM
Sandy Loam	SL	Marl	MARL
Fine Sandy Loam	FSL	Mucky Peat	MPT
Very Fine Sandy Loam	VFSL	Muck	MUCK
Loam	L	Peat	PEAT
Silt Loam	SIL	Cinders	CIND

Phase – Enter an abbreviation for the phase, modifier, or variant of the soil series and describe in the Soil Information Phase Notes section.

Soil Description – Enter Y if a complete and suitable soil description has been made, otherwise enter N.

Slope Percent – Enter percent of slope in whole numbers, 0 to 99. Use 99 for slopes steeper than 99.

Slope Shape – Select the following appropriate code to indicate the slope shape:

Slope Shape	Code
Plane	P
Concave	CV
Convex	CX

Slope Aspect – Select the following appropriate code to indicate the direction the slope faces, to the nearest 45 degrees azimuth.

Direction	Code
North	N
Northeast	NE
East	E
Southeast	SE
South	S
Southwest	SW
West	W
Northwest	NW

Soil Trapped by Windbreak – Indicate in Vital Notes the row or rows affected.

- *Depth* – Enter the inches of soil material trapped by the windbreak at the point of deepest deposition, to the nearest whole inch.

- *Texture* – Select the following appropriate code to indicate the texture of the soil material trapped by the windbreak.

Texture Type	Code	Texture Type	Code
Sand	S	Silt	SI
Coarse Sand	COS	Sandy Clay Loam	SCL
Fine Sand	FS	Clay Loam	CL
Very Fine Sand	VFS	Silty Clay Loam	SICL
Loamy Coarse Sand	LCOS	Sandy Clay	SC
Loamy Sand	LS	Clay	C
Loamy Fine Sand	LFS	Silty Clay	SIC
Loamy Very Fine Sand	LVFS	Fibric Materials	FB
Coarse Sandy Loam	COSL	Hemic Materials	HM
Sandy Loam	SL	Marl	MARL
Fine Sandy Loam	FSL	Mucky Peat	MPT
Very Fine Sandy Loam	VFSL	Muck	MUCK
Loam	L	Peat	PEAT
Silt Loam	SIL	Cinders	CIND

Depth of A Horizon – Enter the depth of the A-Horizon, to the nearest whole inch. The depth should match that shown in the profile description.

Restrictive Layer – Enter the depth to a major change in soil texture, permanent water table, rock, etc., to the nearest whole inch.

Alkaline or Saline –

- *Kind* – Select the appropriate code from the following table:

Kind	Code
None	N
Alkaline	A
Saline	S

- *Degree* – Select the following appropriate code.

Degree	Code
Alkalinity in the soil does not affect windbreak species	N
Alkalinity in the soil affects windbreak species	Y
Salinity <2 mmhos/cm	1
Salinity 2-4 mmhos/cm	2
Salinity 4-8 mmhos/cm	3
Salinity 8-16 mmhos/cm	4
Salinity >16 mmhos/cm	5

(iii) Maintenance and Response

Row Number – Enter the number of the row. Number the rows consecutively from the north or west edge of the windbreak, beginning with the number 1. Where more than one species occurs in a row, add a letter

637-51

starting with A for the first species, B for the second species, etc.

Cultivation – In the *Past* and *Now* fields, select the following code that most nearly indicates the kind of cultivation.

Kind of Cultivation	Code
Excellent - Clean Tilled each Year	E
Good - Clean Tilled Frequently or Partially	G
Fair - Clean Tilled Infrequently or Outside Rows Only	F
Poor - No Cultivation for 3 Years or More	P

Ground Cover – Select the following appropriate code.

Ground Cover	Code
Dense - Heavy Sod	D
Light - Patch Sod or Weeds	L
Sparse - Scattered Grass and Weeds or Clean Tilled	S

Row to Row Spacing (Feet) – Enter the distance between rows of trees or shrubs, to the nearest whole foot.

Tree to Tree Spacing (Feet) – Enter the average distance between individual trees or shrubs within a row, to the nearest whole foot.

Reproduction – Select the following appropriate code that best represents the reproduction of species in the windbreak, either by seed or by vegetative means.

Reproduction	Code
None	N
Light - An Occasional Sprout or Seedling	L
Medium - 6 to 20 Feet Apart	M
Heavy - 5 Feet or Less Apart	H

Damage –

- *Kind* – Select the following appropriate code to indicate the kind of damage to the planting from *Insects, Disease,* and/or *Injury.* Describe the damage in the Maintenance and Response Damage Notes section.

Kind of Damage	Code		
	Insect	Disease	Injury
None	NE	NE	NE
Other	OT	OT	OT
Borer	BR	--	--
Grasshopper	GH	--	--
Oyster Scale	OS	--	--
Tip Moth	TM	--	--
Webworm	WE	--	--
Spruce Mite (Red Spider)	SM	--	--
Elm Leaf Beetle	EB	--	--
Canker	--	CA	--
Cedar Apple Rust	--	CR	--
Fire Blight	--	FR	--
Fungus	--	FN	--

Kind of Damage	Code		
	Insect	Disease	Injury
Needle Rust	--	NR	--
Virus	--	VR	--
X Disease	--	XC	--
Animal (other)	--	--	AL
Chemical	--	--	CM
Deer	--	--	DR
Dieback	--	--	DB
Fire	--	--	FR
Flooding	--	--	FL
Frost Crack	--	--	FC
Grazing	--	--	GR
Hail	--	--	HL
Ice	--	--	IC
Implement	--	--	IP
Rabbit	--	--	RT
Snow	--	--	SN
Sunscald	--	--	SS
Wind	--	--	WN
Winter Kill	--	--	WK

- *Degree* – Select the following appropriate code to indicate the degree of damage to the planting from *Insects, Disease,* and/or *Injury.*

Degree of Damage	Code
Slight - No Appreciable Damage is Apparent	1
Moderate - Apparent Loss of Foliage, Vigor, and Top Growth or General Overall Decline of Species in the Row	2
Severe - Apparent Loss of Species in the Row	3

Rooting Depth – Enter the depth to which the roots of the windbreak species occur in the classes *Many, Common,* and *Few.* See Agriculture Handbook 18, *Soil Survey Manual,* October 1993. Describe any conditions that may influence root development in the soil profile description section.

Vital Notes (Y or N) – Enter a Y (yes) or N (no) to indicate there are additional data in Vital Notes.

(iv) Species Performance
Year Planted – Determine month (if known) and year of the initial windbreak planting. If planting records are not available, bore plants in each row 12 inches (30.5 cm) above the ground. Count the number of annual rings. For deciduous tree and shrub species, the ring count should be subtracted from the year the data were collected to obtain the year the planting was established. For *Juniperus, Pinus, Thuja,* and *Picea* species, add three years to the ring count to obtain the age of the planting.

Row Number – Enter the number of the row. Number the rows consecutively from the north or west edge of the windbreak, beginning with the number 1. Where more than one species occurs in a row, add a letter starting with A for the first species, B for the second species, etc. The row numbers should coincide with the row numbers entered in the Maintenance and Response section.

NSPNS – Enter the national scientific plant name symbol as listed in the Plants database (http://plants.usda.gov).

Age in Years – Enter the number of growing seasons since trees and shrubs were planted. If planting records are not available, bore plants in each row 12 inches above the ground and count the number of annual rings. For deciduous tree and shrub species, the ring count should be subtracted from the year the data were collected to obtain the year the planting was established. For *Juniperus, Pinus, Picea,* and *Thuja* species, add 3 years to the ring count to obtain the age of the planting. In Vital Notes, record the approximate age of the nursery stock when it was planted.

Height in Feet –

- *Average of all Trees* – Enter the average height of the trees recorded in the Tree Measurements section

- *Average of Two Tallest* – Enter the average height of the two tallest trees recorded in the Tree Measurements section, to the nearest whole foot.

Diameter to Whole Inch – Enter the average DBH (diameter at breast height) of the trees recorded in the Tree Measurements section, to the nearest whole inch.

Condition – Select the following appropriate code.

Condition	Code
Good • Leaves (or needles) are normal in color and growth • Small amounts of dead wood (top, branches, and twigs) occur within the live crowns • Evidence of disease, insect, and climatic damage is limited • Little or no evidence of suppression or stagnation	G
Fair • Leaves (or needles) are normal in color and growth • Substantial amounts of dead wood (top, branches, and twigs) occur within the live crowns • Evidence of disease, insect, and climatic damage is obvious • Evidence of suppression or stagnation Current-year growth obviously less than normal	F
Poor • Leaves (or needles) are very abnormal in color and growth • Very large amounts of dead wood (top, branches, and twigs) occur within the live crowns • Evidence of extensive disease, insect, and climatic damage is obvious • Evidence of severe suppression, stagnation, or decadence Current-year growth negligible	P

Suppression –

- *Left* and *Right* – Enter the row number of the trees to the left and right that are suppressing trees in the current row.

- *Degree* – Select the appropriate code from the following table to indicate the degree to which the trees in the rows to the left and right of the current row are being suppressed.

Degree	Code
None - Crown is completely free to develop	0
Slight - Competition for light from the side has caused crown to develop abnormally	1
Moderate - Side competition and/or overtopping has caused abnormal top development and some apparent height loss (up to one-third of total height)	2
Severe - Overtopping has resulted in a serious reduction in height growth (more than one-third of total height)	3

Crown Spread – Enter the average distance that the crown of the species extends to the left and right of the current row, to the nearest whole foot. Left is always the side nearest row 1.

Height to Live Crown – Enter the average height that the species supports live crown to the left and right of the current row, to the nearest whole foot. Left is always the side nearest row 1. If management practices, such as pruning or cultivation, have influenced the crown height, leave these fields blank and explain in the Species Performance Comments section.

Present Survival in Percent – Enter the survival percentage of the species in the row, to the nearest whole percent.

Soil Within Series Range – Enter a Y (yes) or N (no) to indicate if the soil series is within the range of the soil taxonomic unit described for the site.

Tree Measurements – Record measurements for tree species only. Measure six or more trees within the 25-plant segment.

Measured (x) to (x) – Enter the direction in which the trees are numbered, i.e. left to right.

Row/NSPNS – Enter the appropriate row number and the national scientific plant name symbol as listed in the Plants database (http://plants.usda.gov).

Plant Point – The place where trees were originally planted. Begin numbering points as indicated in the "Measured (x) to (x)" field.

637-53

- *Number* – Enter the appropriate number of each tree measured.

- *Height* – Record total height to the nearest foot.

- *Diameter* – Measure the diameter of each single-stemmed tree at breast height (4-1/2 feet) to the nearest inch. For multiple-stemmed trees that branch below the 4-1/2 foot level, measure the diameter of only the largest stem. Show the total number of stems for each multiple-stemmed tree in a small circle immediately above and to the right of the recorded diameter of the largest stem. Measure trees consecutively in the row except those with broken tops, severe diseases, etc.

(c) ESI-Forestland Application

The Ecological Site Inventory application is accessed via the Internet through the ESIS link on the Plants homepage at http://plants.usda.gov. Clicking on the ESIS link (see Figure 637-39) will take you to the ESIS homepage (see Figure 637-40). From the ESIS homepage you can access the Ecological Site Inventory-Forestland application.

(1) Conventions for Using the ESI–Forestland Application

The following conventions are used throughout the ESI–Forestland application:

- Text Boxes – A text box requires that you type in data. The "Soil Series Name" field shown in Figure 637-89 is an example of a text box.

Figure 637-89   Text Box

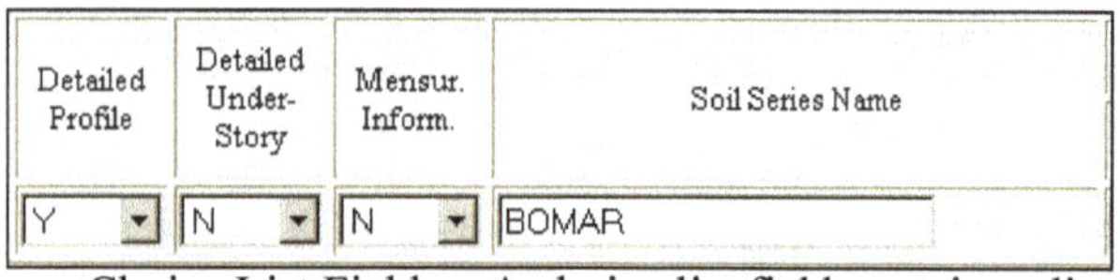

- Choice List Fields – A choice list field contains a list of appropriate items from which to make a selection. Normally, the only data that can be entered in a choice list field are data contained in the choice list. The list of items is viewed from a drop-down choice list.

To select an entry from a drop-down choice list, click on the "down arrow" located on the right edge of the choice list field, then highlight and click on the desired item in the drop-down choice list.

Some choice lists contain more items than can be displayed in the drop-down choice list. To see all of the items, use the scroll bar on the right side of the box to scroll through the items. Figure 637-90 shows an example of a drop-down list of items in a choice list field.

As an alternative to selecting from the drop-down choice list with the mouse, you can also select a choice by placing the cursor in the choice list field and typing the first letter of the first word of your desired choice.

Figure 637-90   Choice List Field

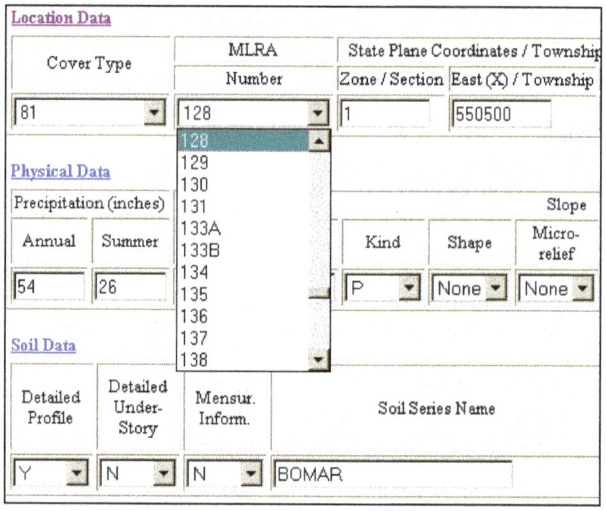

- Hypertext Help – The labels identifying the various tables in the application are also hyperlinks that provide access to help and instructions. To view help and instructions for any table, click on the table label. In most cases these hyperlinks are recognizable as blue, underlined text. When the cursor is placed over the text, the cursor changes to a hand with a pointing finger. The label Location Data in Figure 637-91 is an example of a hypertext help link.

Figure 637-91   Hypertext Help

Action Buttons – The screens used to enter new data or edit existing data are identical, except for the "action" buttons.

When adding new plot data, the following "action" buttons will appear on the screen (see Figure 637-92).

Figure 637-92   Forest Plot Data Entry Screen

- Insert – This button is used to save or insert the data entered on the screen into the database. After clicking on the Insert button, the "action" buttons shown in Figure 637-93 will be displayed.

- Clear – This button will clear all data entered on the screen. It does not delete data that were previously saved or inserted.

- Return/Cancel – This button is used to exit the screen without saving the data currently entered on the screen.

When editing existing plot data, the following action buttons will appear on the screen (see Figure 637-93).

Figure 637-93   Forest Plot Data Edit Screen

- Update – This button is used to replace or update the data previously recorded in the database with the currently edited data.

- Delete – This button is used to delete the entire plot from the database. When you click on the Delete button, the message box shown in Figure 637-94 will appear to confirm the delete action. Click OK to delete the plot. Click Cancel to return to the screen without deleting the plot.

- Reset – This button is similar to the Clear button described above. The difference is that when you click on this button, all the fields on the screen will revert to that data stored in the database, rather than to a blank field.

Figure 637-94   Delete Confirmation Message

- Return/Cancel – This button is used to exit the screen without saving the current edits. When you click on the Return/Cancel button, the message box shown in Figure 637-95 is displayed as a warning that any unsaved data will be lost. Click OK to proceed. Click Cancel to return to the screen without losing edits.

Figure 637-95   Update Reminder

(2) Adding or Editing ESI–Forestland Data
The ESI–Forestland application is designed to allow intermittent data entry. This allows you to enter as little or as much data as is currently available. It is not required to supply all the data associated with an ecological inventory at any one time. The application can be accessed to enter new data or update existing data at any time. The following describes the procedures for accessing the system to add new ESI–Forestland data or edit existing ESI–Forestland data.

(i) Accessing the System
To add or edit ESI–Forestland data, click the Entry/Edit button on the ESI–Forestland homepage (see Figure 637-96) and complete the following on the **Security Login** screen (see Figure 637-97).

Figure 637-96   ESI–Forestland Homepage

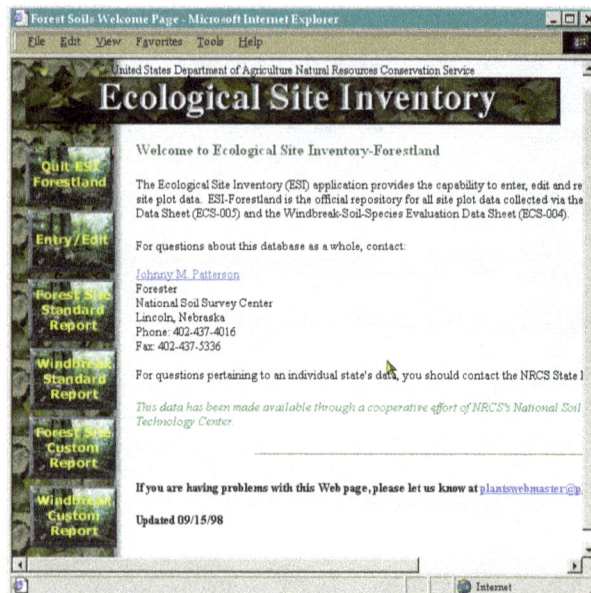

1. Enter Username and Password – Enter your username and password. You cannot add or edit ESI–Forestland data unless you have an authorized username and password. To obtain a username and password, or if you are experiencing difficulty with an existing username and password, contact the individual listed at the bottom of the ESI–Forestland homepage.

Figure 637-97   Security Login Screen

The ESI–Forestland application also allows authorized users to manage their own security information. To change a password, name, and/or telephone number, enter a valid username and password on the **Security Login** screen, then click on the User Profile button . On the subsequent screen (see Figure 637-98) edit the security information as appropriate, then click on the "Update" button to save the changes or click on the "Return" button to leave the screen without saving the edits.

Figure 637-98   User Profile Screen

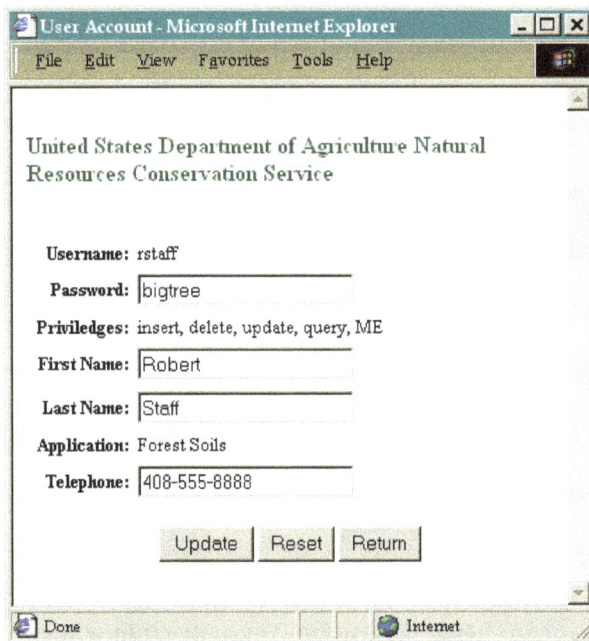

2. Select State, County, and Application – After entering your username and password on the **Security Login** screen , click on the Login button to access the **Data Edit/Entry** screen, as shown in Figure 637-99. Select the desired state from the choice list. Only those states for which you have authorization will be listed in the choice list.

After selecting a state, either type in the desired county name or click on the Browse button and select the desired county from the list of counties in the state **County Choice List** screen (see Figure 637-100).

Choose to edit/enter either forest plot data or windbreak plot data by selecting the appropriate radio button.

3. After completing the above, click on the Search button to display a list of existing forest or

windbreak plots (see Figure 637-101). Use the First, Next, Prev, and Last buttons to navigate through the list.

Figure 637-99   Data Edit/Entry Screen

Figure 637-100   County Choice List Screen

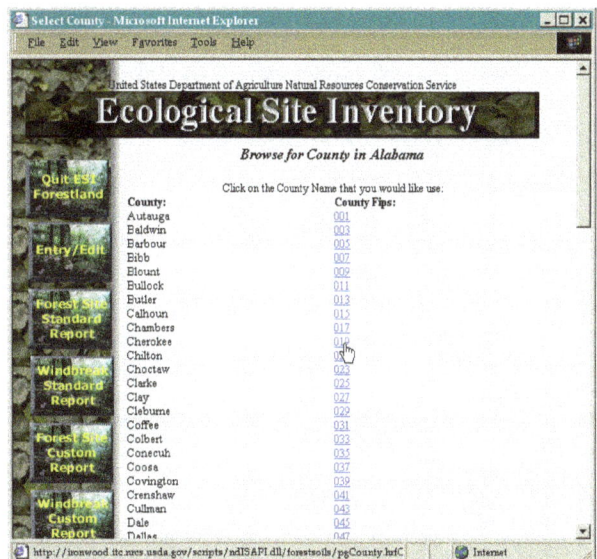

Figure 637-101   Site Selection Screen

637-57

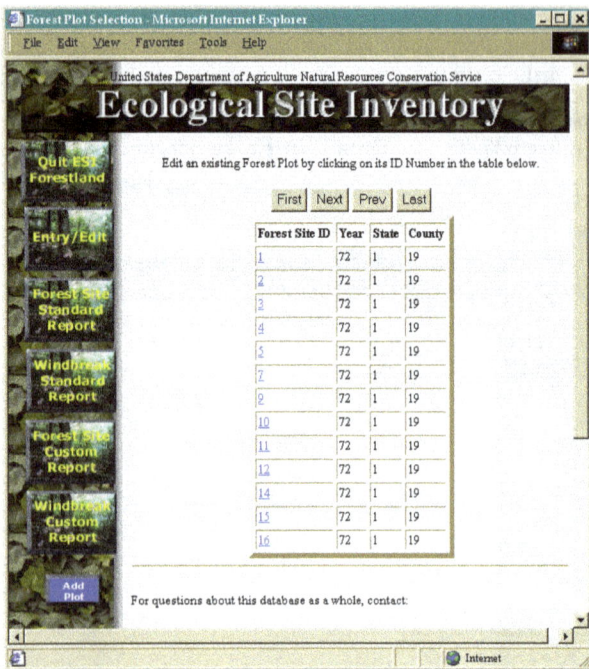

(ii) Forest Plot Data Entry and Edit
After completing the requirements to access the system, you are ready to begin entering data for a new forest plot or editing data for an existing forest plot.

The entry and edit screens are designed to mimic the data entry fields used in the ESI–Forestland Inventory Forest Plot Worksheet. Refer to Part 637.32(a) for instructions and detailed descriptions of the various data fields.

(1) Change Plot ID
To edit a previously entered Forest Site Plot Number, click on the Change Plot ID button (see Figure 637-102) and complete the subsequent entry screen.

(2) Site-Related Data
To enter or edit site-related data (Location Data, Physical Data, Soil Data, and Density Data), simply enter the appropriate data in the text boxes or select the appropriate data from the choice lists.

(3) Vegetative-Related Data
To enter new vegetative-related data (Tree Data, Site Index Data, Canopy Data, Ground Cover Data), click on the appropriate button, located adjacent to the table label (see Figure 637-103) and complete the subsequent screen. **Note, when entering a new forest plot, the**

vegetative-related data fields (Tree Data, Site Index Data, Canopy Data, Ground Cover Data) are not displayed until the new forest plot is saved to the database by clicking on the Update button.

To edit existing vegetative-related data (Tree Data, Site Index Data, Canopy Data, Ground Cover Data), click on the NSPNS for the species you wish to edit (see Figure 637-103) and complete the subsequent screen.

Figure 637-102   Change Forest Plot ID

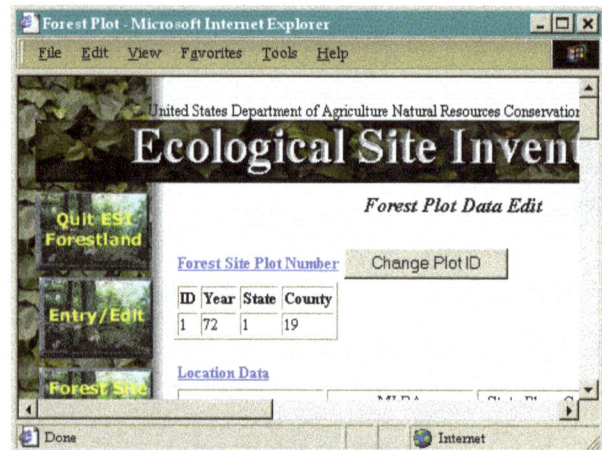

Figure 637-103   Forest Plot Vegetative Information

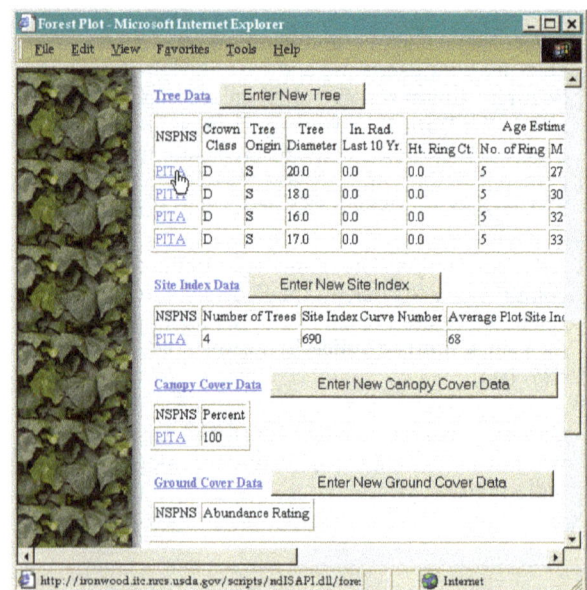

(iii) Windbreak Plot Data Entry and Edit

637-58

After completing the requirements to access the system, you are ready to begin entering data for a new windbreak plot or editing data for an existing windbreak plot.

The entry and edit screens are designed to mimic the data entry fields used in the ESI Windbreak Plot Field Worksheet (see Exhibit 637 - 58). Refer to Part 637.32(b) for instructions and detailed descriptions of the various data fields.

(1) Change Windbreak Plot ID
To edit a previously entered Windbreak Site Plot Number, click on the Change Plot ID button (see Figure 637-104) and complete the subsequent entry screen.

Figure 637-104   Change Windbreak Plot ID

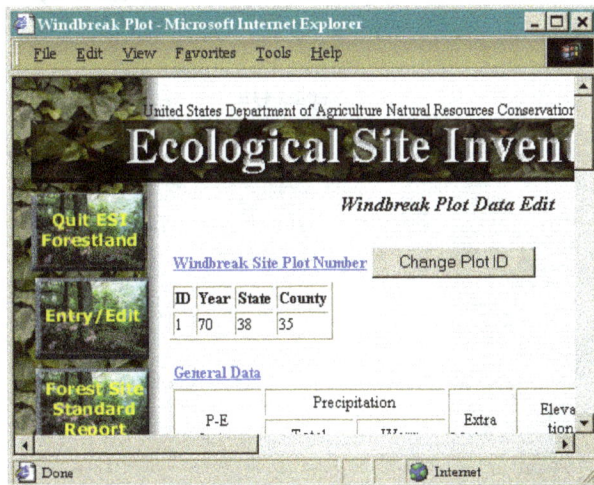

(2) Site-Related Data
To enter or edit site-related data (General Data and Soil Data), simply enter the appropriate data in the text boxes or select the appropriate data from the choice lists.

(3) Vegetative-Related Data
To enter new windbreak row data, click on the Enter New Row button, located adjacent to the table label (see Figure 637-105) and complete the subsequent screen. **Note: When entering a new windbreak plot, the windbreak row table is not displayed until the new windbreak plot is saved to the database by clicking on the Update button.**

To edit existing windbreak row data, click on the Row Position you wish to edit (see Figure 637-105) and complete the subsequent screen.

Figure 637-105   Windbreak Row Data

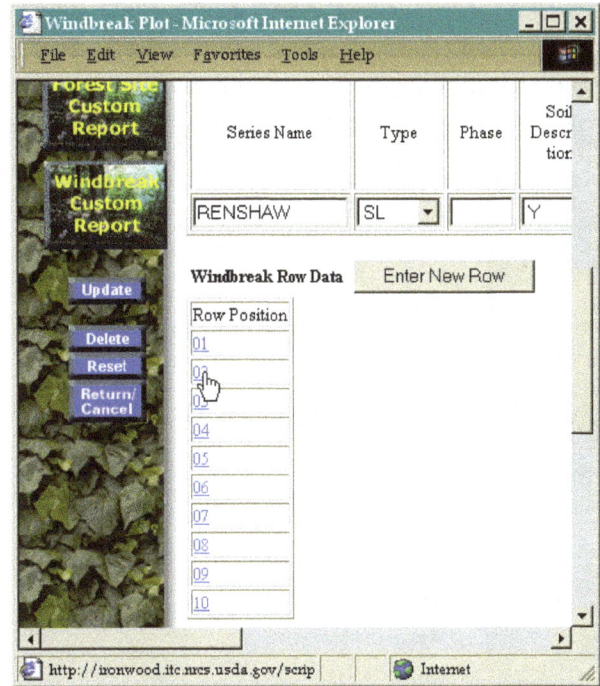

(iv) Producing ESI–Forestland Reports
ESI–Forestland reports are available to anyone with access to the Internet. A password or user name is not required to produce ESI–Forestland reports. Access the reports by clicking one of the report buttons, located on the left side of the ESI–Forestland homepage (see Figure 637-96). The different report formats are detailed below.

(1) Forest Site Standard Reports
The standard report for site index displays selected plot data by soil series and plant species. For each plant species, the report displays the total number of plots, site index mean, site index range, and standard deviation. The plot data displayed can be limited by state, county, MLRA, or soil series. A detailed description of the standard reports is available by clicking on the Report Help hyperlink at the top of the **Standard Report** screen (see Figure 637-106).

To produce a site index standard report, click on the Forest Site Standard Report button and select Site Index on the subsequent screen (see Figure 637-106).

637-59

To produce a ground species standard report, click on the Forest Site Standard Report button and Ground Species on the subsequent screen (see Figure 637-106).

Figure 637-106   Site Index Standard Report

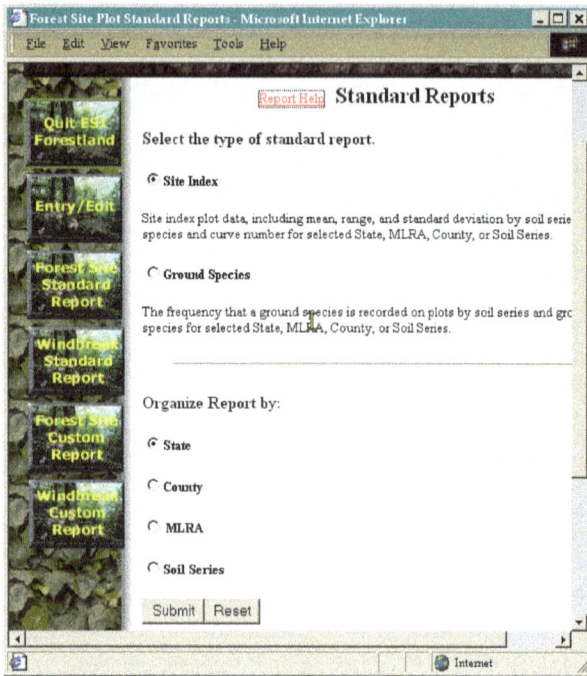

The plot data displayed can be limited by state, county, MLRA, or soil series. To limit the data displayed to certain MLRA's, states, or counties, select the appropriate radio button in the Organize Report By: section, click on the Submit button, and then select one or more MLRA's, states, or counties, from the subsequent choice lists.

To limit the data displayed to certain soil series, select the Soil Series radio button in the Organize Report By: section. On the subsequent screen (see Figure 637-107) select the desired constraint. For example, to limit the report to only those counties where the desired soil series occurs, select County. To display data for the desired soil series regardless of where the soil series occurs, select No Constraints.

Figure 637-107   Soil Series Constrains

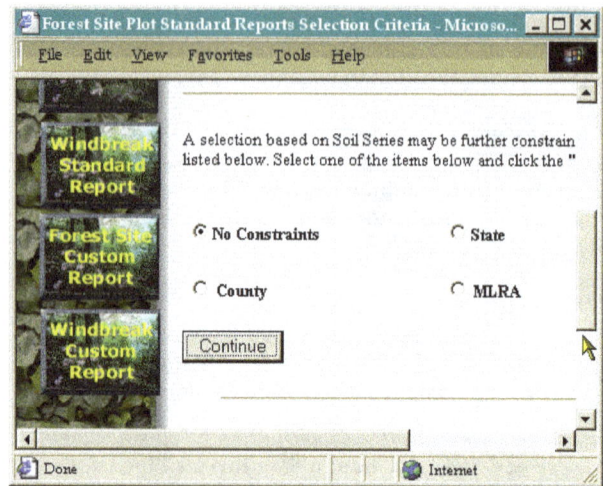

Figure 637-108   Soil Series Query

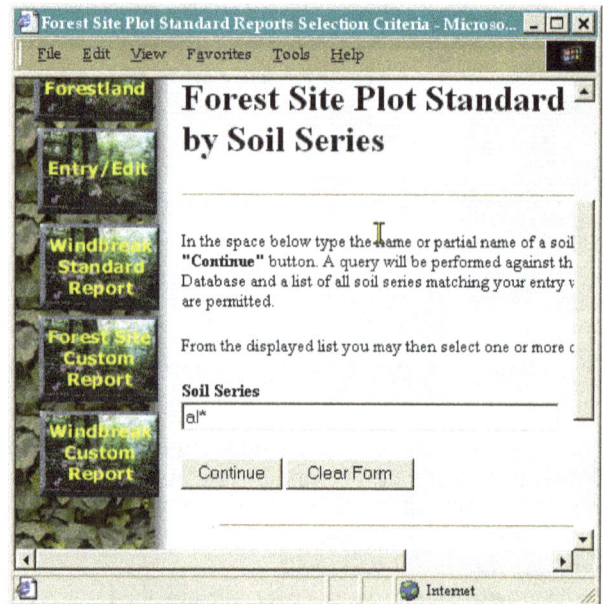

After selecting the desired constraint, click the Submit button. On the subsequent screen (see Figure 637-108), type the name or partial name of a soils series and then click the Continue button. A query will be performed against the database and a list of all soil series matching your entry will be displayed. Wild cards are permitted. An asterisk (*) is used to represent one to many characters, and an underline (_) is used to represent one, and only one, character.

From the subsequent list (see Figure 637-109), select one or more soil series. To select more than one soil series, hold down the shift or control key.

Figure 637-109   Soil Series List

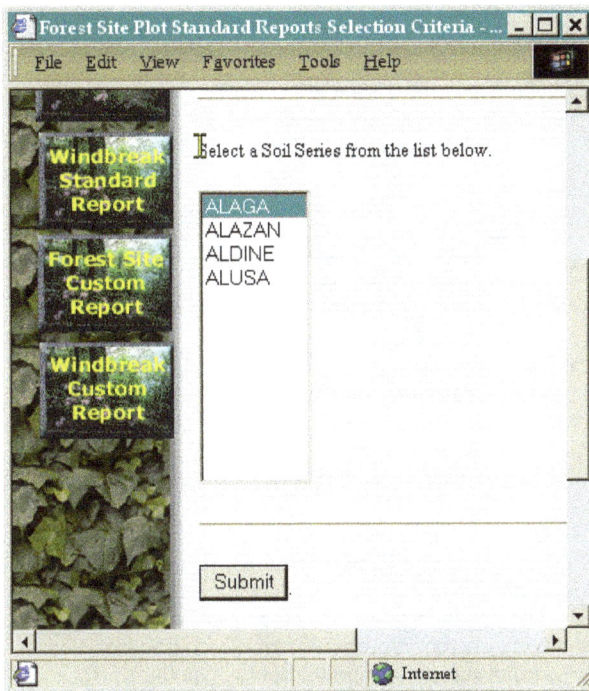

(1) Windbreak Standard Reports
The standard report for windbreaks displays selected plot data by soil series and plant species. The plot data displayed can be limited by state, county, MLRA, or soil series. Refer to the section above on Forest Site Standard Reports for instructions on using the limitation criteria.

(2) Forest Site Custom Report
The forest site custom report function allows users to produce custom reports tailored to their particular interests. To produce a custom report, the user specifies report criteria, selects the columns to be displayed, and specifies the way the report is to be sorted.

First click on the Forest Site Custom Report button located on the left side of the ESI–Forestland homepage (see Figure 637-96). The following describes the basic requirements for producing custom reports. When navigating between the various screens described below, use the <<Back and Next>> buttons located at the

bottom of each screen, rather than using the browser's Back and Forward buttons.

Specify the Report Criteria – Use the screen shown in Figure 637-110 to specify the site-related criteria (State, County, etc.) and select one of the additional criteria buttons (Tree, Site Index, etc.) to specify plant-related criteria. This feature allows you to limit the report to only the data that match the criteria you specify.

Figure 637-110   Criteria Screen

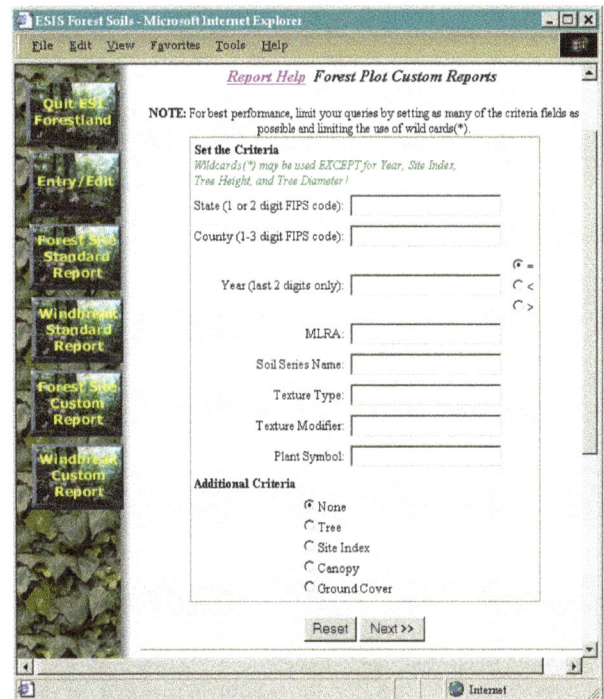

Some fields in the database (Texture Type, Texture Modifier, etc.) are stored as codes – FSL is the code for a fine sandy loam texture.

To determine the proper codes, refer to the Help Screen Links located in the on-line help documents. Access the help document by clicking on the Report Help hyperlink at the top of each custom report screen (see Figure 637-110).

Do not enter leading zeros (0) when entering the State or County FIPS code. For example, if the state FIPS code is 005, enter 5.

Specify Additional Criteria and Select Columns – Use the screen shown in Figure 637-111 to select the column

637-61

names that correspond to the data you want to view. If you selected one of the additional criteria buttons on the initial criteria screen, you can specify further criteria relative to your selection on this screen. For example, if the additional criterion button Site Index was selected on the initial criteria screen, the subsequent screen would look like the screen shown in Figure 637-111.

Figure 637-111   Column Select Screen

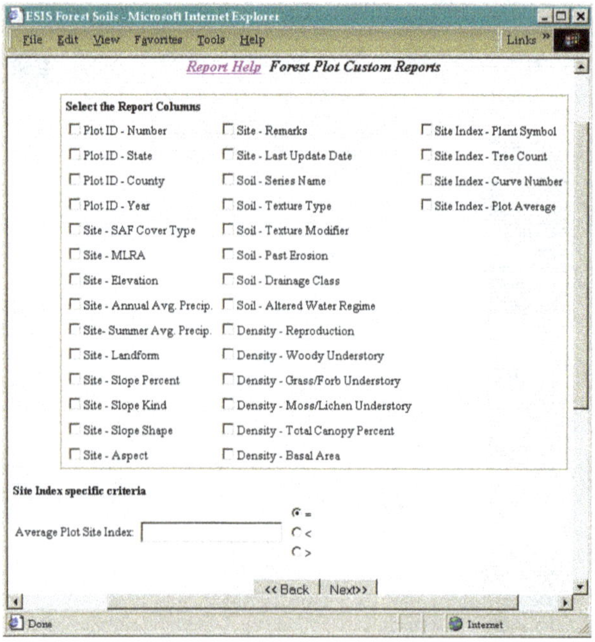

Select the Sorting Order – On the screen shown in Figure 637-112, select the way you want the data sorted in the report. The data in the report may be sorted by up to three of the columns previously selected. Unselected columns are not available for sorting. On this screen you can also select the report format. Select either of the delimited formats if you wish to import the data into other programs, such as a spreadsheet or database. Choose the HTML format to view the data with your browser.

(3) Windbreak Custom Report
The windbreak custom report functionality allows users to produce custom reports tailored to their particular interests. To produce a custom report the user specifies report criteria, selects the columns to be displayed, and specifies the way the report is to be sorted.

To produce a windbreak custom report, click on the Windbreak Custom Report button, , located on the left side of the ESI–Forestland homepage (see Figure 637 - 96). The following describes the basic requirements for producing custom reports. When navigating between the various screens described below, use the <<Back and Next>> buttons, located at the bottom of each screen rather than using the browser's Back and Forward buttons.

Specify the Report Criteria — Use the screen shown in Figure 637-113 to specify the site-related criteria (State, County, Texture Type, etc.). This feature allows you to limit the report to only that data that matches the criteria you specify.

Some fields in the database, like Texture Type, are stored as codes. For example – FSL is the code for a fine sandy loam texture type.

To determine the proper codes, refer to the Help Screen Links located in the on-line help documents. Access the help document by clicking on the Report Help hyperlink at the top of each custom report screen (see Figure 637-114).

Figure 637-112   Sort Screen

Do not enter leading zeros (0) when entering the State or County FIPS code. For example, if the state FIPS code is 005, enter 5.

Figure 637-113   Windbreak Criteria Screen

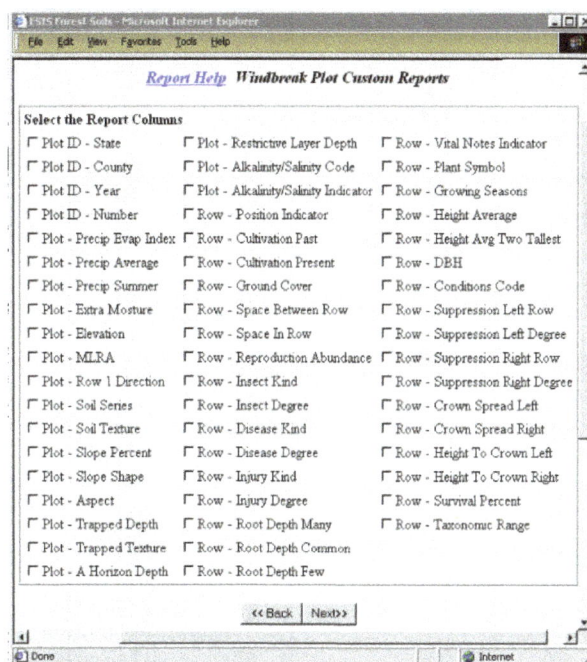

Figure 637-114   Windbreak Column Select Screen

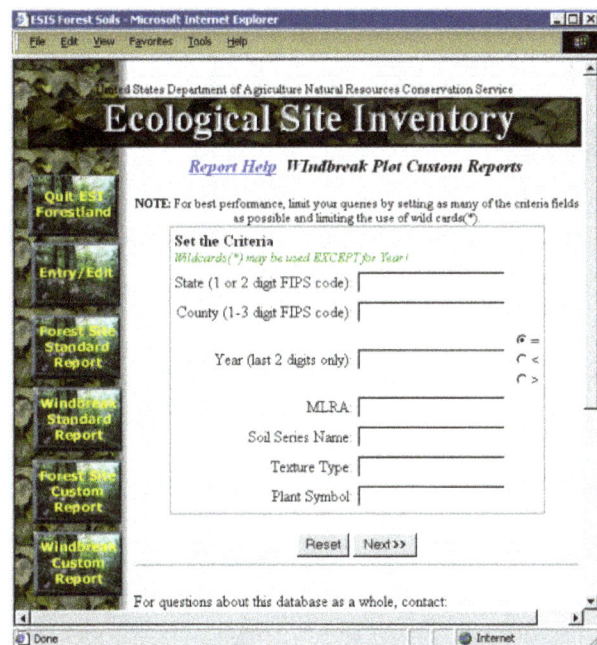

Select Columns — Use the screen shown in Figure 637-114 to select the column names that correspond to the data you want to view.

Select the Sort Order — On the screen shown in Figure 637-115, select the way you want the data sorted in the report. The data in the report may be sorted by up to three of the columns previously selected. Unselected columns are not available for sorting. On this screen you can also select the report format. Select either of the delimited formats if you wish to import the data into other programs, such as a spreadsheet or database. Chose the HTML format to view the data with your browser.

Figure 637-15   Windbreak Sort Screen

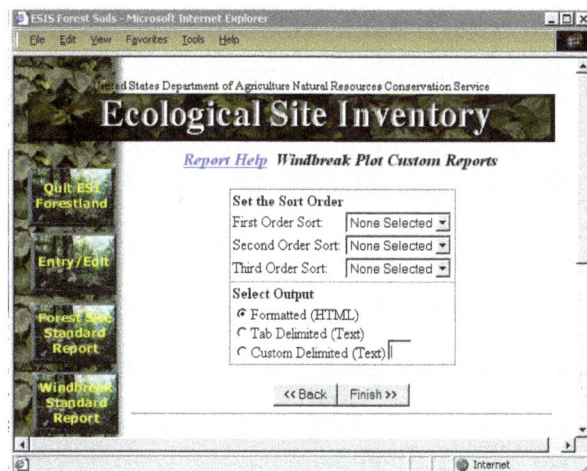

637-63

Exhibit 637-1   CMAI Annual Growth - *Abies concolor* (white fir)

Site Index	Ft³/Ac	Site Index	Ft³/Ac
30	51	70	163
31	52	71	166
32	54	72	170
33	55	73	173
34	56	74	176
35	57	75	179
36	59	76	183
37	60	77	186
38	61	78	189
39	63	79	193
40	64	80	196
41	67	81	198
42	69	82	200
43	72	83	202
44	75	84	204
45	77	85	206
46	80	86	209
47	83	87	211
48	86	88	213
49	88	89	215
50	91	90	217
51	95		
52	98		
53	102		
54	106		
55	109		
56	113		
57	117		
58	121		
59	124		
60	128		
61	131		
62	135		
63	138		
64	142		
65	145		
66	149		
67	152		
68	156		
69	159		

Exhibit 637-2   CMAI Annual Growth - *Abies magnifica* (California red fir)

Site Index	Ft³/Ac
20	76
21	79
22	82
23	84
24	87
25	90
26	93
27	96
28	98
29	101
30	104
31	107
32	110
33	113
34	116
35	119
36	123
37	126
38	129
39	132
40	135
41	139
42	142
43	146
44	149
45	153
46	157
47	160
48	164
49	167
50	171
51	175
52	180
53	184
54	188
55	192
56	197
57	201
58	205
59	210

Site Index	Ft³/Ac
60	214

Exhibit 637-3   CMAI Annual Growth - *Abies balsamea* (balsam fir)

Site Index	Ft³/Ac
34	58
35	60
36	62
37	65
38	67
39	69
40	71
41	74
42	76
43	78
44	80
45	83
46	86
47	88
48	90
49	93
50	96
51	98
52	100
53	102
54	105
55	107
56	109
57	111
58	113
59	116
60	118
61	120
62	122
63	124
64	127
65	129
66	131
67	133
68	135
69	138
70	140

Exhibit 637-4   CMAI Annual Growth - *Abies lasiocarpa* (subalpine fir)

Site Index	Ft³/Ac	Site Index	Ft³/Ac
40	28	80	76
41	29	81	77
42	30	82	79
43	59	83	80
44	31	84	82
45	32	85	83
46	33	86	85
47	34	87	86
48	35	88	88
49	36	89	89
50	37	90	91
51	39	91	92
52	40	92	94
53	41	93	96
54	45	94	97
55	43	95	99
56	44	96	100
57	45	97	102
58	46	98	104
59	48	99	106
60	49	100	109
61	50		
62	51		
63	52		
64	53		
65	55		
66	57		
67	59		
68	60		
69	61		
70	63		
71	64		
72	65		
73	66		
74	68		
75	69		
76	70		
77	72		
78	73		
79	74		

Exhibit 637-5   CMAI Annual Growth - *Acer saccharinum* (silver maple)

Site Index	Ft³/Ac
60	16
61	17
62	18
63	19
64	20
65	20
66	21
67	22
68	23
69	24
70	25
71	26
72	27
73	28
74	29
75	29
76	30
77	31
78	32
79	33
80	34
81	35
82	36
83	36
84	37
85	38
86	39
87	40
88	40
89	41
90	42
91	43
92	44
93	45
94	46
95	46
96	47
97	48
98	49

Site Index	Ft³/Ac
99	50
100	51

Exhibit 637-6   CMAI Annual Growth - *Alnus rubra* (red alder)

Site Index	Ft³/Ac	Site Index	Ft³/Ac
60	50	99	116
61	51	100	118
62	53	101	120
63	55	102	121
64	56	103	123
65	58	104	125
66	60	105	127
67	61	106	128
68	53	107	130
69	65	108	132
70	66	109	133
71	58	110	135
72	70	111	137
73	72	112	138
74	73	113	140
75	75	114	142
76	77	115	144
77	79	116	145
78	80	117	147
79	82	118	149
80	84	119	150
81	86	120	152
82	87		
83	89		
84	91		
85	92		
86	94		
87	96		
88	97		
89	99		
90	101		
91	102		
92	104		
93	106		
94	108		
95	109		
96	111		
97	113		
98	115		

637-69

Exhibit 637-7   CMAI Annual Growth - *Betula papyrifera* (paper birch)

Site Index	Ft³/Ac
40	35
41	36
42	37
43	37
44	38
45	39
46	40
47	41
48	41
49	42
50	43
51	44
52	45
53	46
54	47
55	48
56	50
57	51
58	52
59	53
60	54

Exhibit 637-8   CMAI Annual Growth - *Betula papyrifera* (paper birch) - Alaska

Site Index	Ft³/Ac
35	12
36	13
37	14
38	14
39	15
40	16
41	17
42	18
43	18
44	19
45	20
46	21
47	22
48	24
49	25
50	26
51	28
52	29
53	30
54	31
55	32
56	34
57	35
58	37
59	38
60	40
61	41
62	43
63	44
64	46
65	48

Exhibit 637-9   CMAI Annual Growth - *Fraxinus pennsylvanica* (green ash)

Site Index	Ft³/Ac		Site Index	Ft³/Ac
65	25		105	91
66	27		106	92
67	28		107	94
68	30		108	96
69	32		109	97
70	33		110	99
71	35			
72	36			
73	38			
74	40			
75	41			
76	43			
77	44			
78	46			
79	48			
80	49			
81	51			
82	52			
83	54			
84	56			
85	57			
86	59			
87	60			
88	62			
89	64			
90	65			
91	67			
92	69			
93	70			
94	72			
95	74			
96	75			
97	77			
98	79			
99	80			
100	82			
101	84			
102	85			
103	87			
104	89			

Exhibit 637-10 CMAI Annual Growth - *Juniperus L.* (juniper) - Southwest

Site Index	Ft³/Ac
20	1.3
25	1.7
60	2.2
35	2.7
40	3.3
45	3.8
50	4.6
55	5.2
60	6.1
65	6.7
70	7.8
75	8.5
80	9.8
85	10.6
90	12.3
95	13.1
100	15.2
105	16.0
110	18.3
115	19.2
120	22.1
125	23.3
130	25.0
135	26.7
140	28.4
145	30.0
150	31.7
155	33.4
160	35.1

Exhibit 637-11   CMAI Annual Growth - *Juniperus occidentalis* (western juniper)

Site Index	Ft^3/Ac
15-20	±15
20-30	±30
30-35	±45

Exhibit 637-12   CMAI Annual Growth - *Larix occidentalis* (western larch)

Site Index	Ft³/Ac
30	41
31	42
32	43
33	44
34	45
35	46
36	47
37	48
38	49
39	50
40	51
41	52
42	53
43	55
44	57
45	58
46	60
47	61
48	63
49	65
50	67
51	69
52	71
53	73
54	75
55	78
56	80
57	82
58	85
59	87
60	90
61	92
62	94
63	97
64	99
65	102
66	105
67	108
68	110
69	113

Site Index	Ft³/Ac
70	115
71	117
72	120
73	123
74	126
75	129
76	131
77	133
78	136
79	139
80	141
81	144
82	147
83	150
84	153
85	155
86	158
87	160
88	162
89	165
90	168

Exhibit 637-13   CMAI Annual Growth - *Larix laricina* (tamarack)

Site Index	Ft³/Ac
30	18
31	19
32	20
33	21
34	22
35	23
36	23
37	24
38	25
39	26
40	27
41	29
42	30
43	32
44	33
45	35
46	36
47	38
48	39
49	41
50	42
51	44
52	45
53	47
54	49
55	50
56	52
57	54
58	56
59	57
60	59
61	61
62	62
63	64
64	65
65	67
66	69
67	70
68	72
69	73

Site Index	Ft³/Ac
70	75

Exhibit 637-14   CMAI Annual Growth - *Liquidambar styraciflua* (sweetgum)

Site Index	Ft³/Ac		Site Index	Ft³/Ac
70	57		110	173
71	59		111	176
72	61		112	180
73	64		113	184
74	66		114	187
75	68		115	190
76	70		116	194
77	72		117	198
78	75		118	201
79	77		119	204
80	79		120	208
81	82			
82	84			
83	87			
84	90			
85	93			
86	95			
87	98			
88	101			
89	103			
90	106			
91	109			
92	112			
93	116			
94	119			
95	122			
96	125			
97	128			
98	132			
99	135			
100	138			
101	142			
102	145			
103	148			
104	152			
105	156			
106	159			
107	162			
108	166			
109	170			

Exhibit 637-15   CMAI Annual Growth - *Liriodendron tulipifera* (tuliptree)

Site Index	Ft³/Ac		Site Index	Ft³/Ac
60	37		100	107
61	39		101	109
62	40		102	110
63	42		103	112
64	44		104	114
65	45		105	115
66	47		106	117
67	49		107	119
68	51		108	121
69	52		109	122
70	54		110	124
71	56		111	126
72	57		112	127
73	59		113	128
74	61		114	130
75	62		115	132
76	64		116	133
77	66		117	134
78	68		118	136
79	69		119	138
80	71		120	139
81	73			
82	75			
83	77			
84	79			
85	81			
86	82			
87	84			
88	86			
89	88			
90	90			
91	92			
92	93			
93	95			
94	97			
95	98			
96	100			
97	102			
98	104			
99	105			

Exhibit 637-16   CMAI Annual Growth - *Nyssa aquatica* (water tupelo)

Site Index	Ft³/Ac	Site Index	Ft³/Ac
50	70	90	141
51	71	91	145
52	72	92	148
53	73	93	152
54	74	94	156
55	76	95	160
56	77	96	163
57	78	97	167
58	79	98	171
59	80	99	174
60	81	100	178
61	82		
62	84		
63	85		
64	87		
65	88		
66	89		
67	91		
68	92		
69	94		
70	95		
71	97		
72	99		
73	101		
74	103		
75	105		
76	106		
77	108		
78	110		
79	112		
80	114		
81	117		
82	119		
83	122		
84	125		
85	128		
86	130		
87	133		
88	136		
89	138		

Exhibit 637-17   CMAI Annual Growth - *Nyssa biflora* (swamp tupelo)

Site Index	Ft³/Ac	Site Index	Ft³/Ac
50	70	90	141
51	71	91	145
52	72	92	148
53	73	93	152
54	74	94	156
55	76	95	160
56	77	96	163
57	78	97	167
58	79	98	171
59	80	99	174
60	81	100	178
61	82		
62	84		
63	85		
64	87		
65	88		
66	89		
67	91		
68	92		
69	94		
70	95		
71	97		
72	99		
73	101		
74	103		
75	105		
76	106		
77	108		
78	110		
79	112		
80	114		
81	117		
82	119		
83	122		
84	125		
85	128		
86	130		
87	133		
88	136		
89	138		

Exhibit 637-18   CMAI Annual Growth - *Picea engelmannii* (Engelmann's spruce)

Site Index	Ft³/Ac	Site Index	Ft³/Ac
40	28	80	76
41	29	81	77
42	30	82	79
43	59	83	80
44	31	84	82
45	32	85	83
46	33	86	85
47	34	87	86
48	35	88	88
49	36	89	89
50	37	90	91
51	39	91	92
52	40	92	94
53	41	93	96
54	45	94	97
55	43	95	99
56	44	96	100
57	45	97	102
58	46	98	104
59	48	99	106
60	49	100	109
61	50		
62	51		
63	52		
64	53		
65	55		
66	57		
67	59		
68	60		
69	61		
70	63		
71	64		
72	65		
73	66		
74	68		
75	69		
76	70		
77	72		
78	73		
79	74		

Exhibit 637-19   CMAI Annual Growth - *Picea glauca* (white spruce)

Site Index	Ft³/Ac		Site Index	Ft³/Ac
50	12		90	41
51	12		91	42
52	13		92	43
53	13		93	44
54	14		94	45
55	15		95	46
56	15		96	47
57	16		97	48
58	17		98	49
59	17		99	50
60	18		100	51
61	18			
62	19			
63	20			
64	20			
65	21			
66	22			
67	22			
68	23			
69	24			
70	25			
71	25			
72	26			
73	27			
74	28			
75	28			
76	29			
77	30			
78	31			
79	31			
80	32			
81	32			
82	33			
83	34			
84	35			
85	36			
86	37			
87	38			
88	39			
89	40			

Exhibit 637-20   CMAI Annual Growth - *Picea pungens* (Colorado blue spruce)

Site Index	Ft³/Ac
15	23.0
16	24.2
17	25.4
18	26.6
19	27.8
20	29.0
21	30.2
22	31.4
23	32.6
24	33.8
25	35.0
26	36.1
27	37.2
28	38.3
29	39.4
30	40.5
31	41.6
32	42.7
33	43.8
34	44.9
35	46.0

Exhibit 637-21   CMAI Annual Growth - *Picea rubens* (red spruce)

Site Index	Ft³/Ac
30	52
31	56
32	59
33	62
34	66
35	70
36	73
37	76
38	80
39	84
40	87
41	89
42	91
43	94
44	96
45	98
46	100
47	102
48	105
49	107
50	109
51	112
52	115
53	118
54	121
55	123
56	126
57	129
58	132
59	135
60	138
61	141
62	144
63	147
64	150
65	152
66	155
67	158
68	161
69	164

Site Index	Ft³/Ac
70	167

Exhibit 637-22   CMAI Annual Growth - *Picea sitchensis* (Sitka spruce)

Site Index	Ft³/Ac	Site Index	Ft³/Ac	Site Index	Ft³/Ac	Site Index	Ft³/Ac
60	66	96	113	132	181	168	253
61	67	97	115	133	183	169	255
62	68	98	116	134	185	170	256
63	69	99	118	135	188	171	258
64	71	100	120	136	190	172	260
65	72	101	122	137	193	173	261
66	72	102	123	138	195	174	262
67	73	103	125	139	198	175	264
68	74	104	126	140	200	176	265
69	75	105	128	141	203	177	267
70	76	106	130	142	206	178	268
71	77	107	132	143	208	179	270
72	78	108	134	144	210	180	271
73	80	109	136	145	212	181	273
74	81	110	138	146	214	182	274
75	82	111	140	147	216	183	276
76	83	112	142	148	218	184	277
77	84	113	143	149	220	185	279
78	86	114	145	150	222	186	280
79	87	115	147	151	224	187	282
80	88	116	149	152	226	188	283
81	89	117	151	153	227	189	285
82	91	118	153	154	229	190	286
83	92	119	154	155	231	191	287
84	93	120	156	156	233	192	289
85	95	121	158	157	235	193	290
86	96	122	160	158	237	194	292
87	98	123	162	159	239	195	293
88	99	124	164	160	240	196	295
89	101	125	166	161	242	197	296
90	102	126	168	162	244	198	297
91	105	127	170	163	245	199	299
92	107	128	172	164	247	200	300

Site Index	Ft³/Ac	Site Index	Ft³/Ac	Site Index	Ft³/Ac
93	108	129	174	165	248
94	110	130	176	166	250
95	112	131	179	167	252

637-85

Exhibit 637-23   CMAI Annual Growth - *Pinus L.* (pinyon pine) – Southwest

Site Index	Ft³/Ac
20	1.3
25	1.7
60	2.2
35	2.7
40	3.3
45	3.8
50	4.6
55	5.2
60	6.1
65	6.7
70	7.8
75	8.5
80	9.8
85	10.6
90	12.3
95	13.1
100	15.2
105	16.0
110	18.3
115	19.2
120	22.1
125	23.3
130	25.0
135	26.7
140	28.4
145	30.0
150	31.7
155	33.4
160	35.1

Exhibit 637-24   CMAI Annual Growth - *Pinus banksiana* (jack pine)

Site Index	Ft³/Ac
40	56
41	58
42	59
43	61
44	63
45	64
46	66
47	68
48	70
49	71
50	73
51	75
52	76
53	78
54	80
55	81
56	83
57	84
58	86
59	88
60	89
61	91
62	92
63	94
64	96
65	97
66	99

Exhibit 637-25   CMAI Annual Growth - *Pinus clausa* (sand pine)

Site Index	Ft^3/Ac
40	20
41	21
42	22
43	23
44	24
45	25
46	26
47	27
48	28
49	29
50	30
51	31
52	32
53	33
54	34
55	35
56	37
57	38
58	39
59	40
60	42
61	43
62	44
63	46
64	47
65	49
66	50
67	52
68	53
69	54
70	56
71	57
72	59
73	60
74	61
75	63
76	64
77	66
78	67
79	69

Site Index	Ft^3/Ac
80	71
81	72

Exhibit 637-26   CMAI Annual Growth - *Pinus contorta* (lodgepole pine)

Site Index	Ft³/Ac	Site Index	Ft³/Ac
45	36	85	74
46	37	86	75
47	38	87	76
48	39	88	77
49	40	89	78
50	41	90	79
51	41	91	80
52	45	92	81
53	43	93	82
54	44	94	83
55	45	95	84
56	46	96	85
57	47	97	86
58	48	98	87
59	49	99	88
60	50	100	89
61	50	101	90
62	51	102	91
63	52	103	92
64	53	104	93
65	54	105	94
66	55	106	95
67	56	107	96
68	57	108	97
69	58	109	98
70	59	110	99
71	60	111	100
72	61	112	101
73	65	113	102
74	63	114	103
75	65	115	104
76	65	116	105
77	66	117	106
78	67	118	107
79	68	119	108
80	69	120	109
81	70		
82	71		
83	72		
84	73		

Exhibit 637-27   CMAI Annual Growth - *Pinus echinata* (shortleaf pine)

Site Index	Ft³/Ac		Site Index	Ft³/Ac
40	47		80	130
41	49		81	132
42	51		82	134
43	53		83	136
44	55		84	138
45	57		85	140
46	60		86	142
47	62		87	144
48	64		88	148
49	66		89	150
50	68		90	152
51	70		91	154
52	72		92	157
53	74		93	159
54	76		94	161
55	78		95	162
56	80		96	163
57	82		97	165
58	84		98	168
59	86		99	170
60	88		100	172
61	90			
62	92			
63	95			
64	97			
65	99			
66	101			
67	103			
68	106			
69	108			
70	110			
71	112			
72	114			
73	116			
74	118			
75	120			
76	122			
77	124			
78	126			
79	128			

Exhibit 637-28   CMAI Annual Growth - *Pinus elliottii* (slash pine)

Site Index	Ft³/Ac
60	93
61	96
62	98
63	100
64	104
65	107
66	109
67	112
68	114
69	117
70	120
71	122
72	125
73	127
74	129
75	132
76	134
77	136
78	138
79	141
80	143
81	145
82	147
83	149
84	151
85	153
86	155
87	157
88	159
89	161
90	163
91	165
92	167
93	169
94	171
95	172
96	174
97	177
98	179
99	181

Site Index	Ft³/Ac
100	183
101	185
102	186
103	187
104	189
105	191

Exhibit 637-29 CMAI Annual Growth - *Pinus elliottii densa* (South Florida slash pine)

Site Index	Ft³/Ac
30	41
31	42
32	43
33	44
34	45
35	46
36	47
37	48
38	49
39	50
40	51
41	52
42	54
43	55
44	57
45	59
46	61
47	63
48	64
49	66
50	68
51	71
52	74
53	77
54	80
55	83
56	87
57	91
58	95
59	98
60	102
61	108
62	116
63	125
64	133
65	141
66	149
67	157
68	166
69	174

Site Index	Ft³/Ac
70	182

637-92

Exhibit 637-30 CMAI Annual Growth - *Pinus monticola* (western white pine)

Site Index	Ft³/Ac	Site Index	Ft³/Ac
27	62	67	130
28	64	68	132
29	65	69	133
30	67	70	135
31	69	71	137
32	70	72	139
33	72	73	141
34	74	74	146
35	75	75	144
36	77	76	146
37	79	77	148
38	81	78	150
39	82	79	152
40	84	80	154
41	86	81	156
42	87	82	158
43	89	83	160
44	91	84	162
45	92	85	164
46	94	86	165
47	96	87	167
48	98		
49	99		
50	101		
51	103		
52	104		
53	106		
54	108		
55	110		
56	111		
57	113		
58	115		
59	116		
60	118		
61	120		
62	121		
63	123		
64	125		
65	126		
66	128		

Exhibit 637-31   CMAI Annual Growth - *Pinus palustris* (longleaf pine)

Site Index	Ft³/Ac	Site Index	Ft³/Ac
40	19	80	100
41	21	81	103
42	22	82	106
43	24	83	107
44	26	84	110
45	27	85	112
46	29	86	114
47	31	87	117
48	33	88	119
49	34	89	122
50	36	90	124
51	38	91	126
52	40	92	127
53	42	93	129
54	43	94	130
55	45	95	133
56	47	96	134
57	50	97	137
58	52	98	139
59	54	99	140
60	56	100	143
61	57	101	144
62	60	102	146
63	63	103	147
64	65	104	149
65	67	105	151
66	70	106	151
67	72	107	153
68	74	108	154
69	77	109	156
70	79	110	157
71	81		
72	83		
73	86		
74	88		
75	90		
76	92		
77	94		
78	97		
79	99		

Exhibit 637-32   CMAI Annual Growth - *Pinus ponderosa* (ponderosa pine)

Site Index	Ft³/Ac		Site Index	Ft³/Ac		Site Index	Ft³/Ac		Site Index	Ft³/Ac
40	30		80	69		120	141		160	234
41	31		81	70		121	144			
42	31		82	72		122	146			
43	32		83	74		123	149			
44	33		84	75		124	151			
45	34		85	77		125	154			
46	34		86	78		126	156			
47	35		87	80		127	159			
48	36		88	82		128	161			
49	37		89	83		129	164			
50	38		90	85		130	166			
51	38		91	87		131	168			
52	39		92	88		132	170			
53	40		93	90		133	173			
54	41		94	92		134	175			
55	42		95	94		135	177			
56	42		96	96		136	179			
57	43		97	97		137	181			
58	44		98	99		138	183			
59	45		99	101		139	185			
60	46		100	102		140	188			
61	47		101	104		141	190			
62	48		102	106		142	192			
63	49		103	108		143	194			
64	50		104	110		144	197			
65	50		105	112		145	199			
66	51		106	114		146	201			
67	52		107	116		147	203			
68	53		108	118		148	205			
69	54		109	120		149	208			
70	55		110	122		150	210			
71	56		111	124		151	212			
72	58		112	126		152	215			
73	59		113	128		153	217			
74	60		114	130		154	220			
75	62		115	132		155	222			
76	63		116	134		156	224			
77	64		117	136		157	227			
78	65		118	137		158	229			
79	67		119	139		159	232			

Exhibit 637-33   CMAI Annual Growth - *Pinus resinosa* (red pine)

Site Index	Ft³/Ac
45	54
46	56
47	59
48	61
49	63
50	65
51	68
52	70
53	72
54	75
55	77
56	80
57	83
58	86
59	89
60	92
61	95
62	98
63	101
64	104
65	107

Exhibit 637-34   CMAI Annual Growth - *Pinus serotina* (pond pine)

Site Index	Ft³/Ac
40	18
41	19
42	20
43	21
44	22
45	23
46	24
47	25
48	26
49	27
50	28
51	29
52	30
53	31
54	32
55	33
56	34
57	35
58	37
59	38
60	39
61	40
62	42
63	43
64	44
65	46
66	47
67	48
68	51
69	52
70	53
71	55
72	56
73	57
74	59
75	61
76	62
77	63
78	65

Site Index	Ft³/Ac
79	66
80	68
81	69
82	71
83	72
84	74
85	75
86	77
87	78
88	80
89	81
90	82

Exhibit 637-35 CMAI Annual Growth - *Pinus strobus* (eastern white pine) - WV, MD, DE, VA

Site Index	Ft³/Ac
50	72
51	74
52	77
53	80
54	82
55	84
56	87
57	90
58	92
59	94
60	97
61	99
62	102
63	104
64	107
65	109
66	111
67	114
68	116
69	119
70	121
71	123
72	126
73	128
74	130
75	132
76	135
77	137
78	139
79	142
80	144
81	146
82	148
83	151
84	153
85	155
86	157
87	159
88	162
89	164

Site Index	Ft³/Ac
90	166
91	168
92	170
93	172
94	174
95	176
96	178
97	180
98	182
99	184
100	186

Exhibit 637-36   CMAI Annual Growth - *Pinus strobus* (eastern white pine)

Site Index	Ft³/Ac
45	75
46	78
47	81
48	84
49	87
50	90
51	93
52	96
53	99
54	102
55	106
56	109
57	112
58	115
59	118
60	121
61	124
62	127
63	130
64	133
65	136
66	139
67	142
68	145
69	148
70	151
71	154
72	157
73	159
74	162
75	164

Exhibit 637-37   CMAI Annual Growth - *Pinus strobus* (eastern white pine) - PA, NJ, NY, and New England

Site Index	Ft³/Ac
40	62
41	64
42	66
43	68
44	70
45	72
46	73
47	75
48	77
49	79
50	81
51	83
52	85
53	87
54	89
55	92
56	94
57	96
58	98
59	100
60	102
61	104
62	107
63	110
64	112
65	114
66	117
67	120
68	122
69	125
70	127
71	129
72	131
73	133
74	135
75	137
76	139

Site Index	Ft³/Ac
77	141
78	143
79	145
80	147

Source: Frothingham, E.H. 1914. **White pine under forest management**. USDA, Forest Service Bulletin 13.

Exhibit 637-38   CMAI Annual Growth - *Pinus taeda* (loblolly pine)

Site Index	Ft³/Ac	Site Index	Ft³/Ac
60	76	100	154
61	78	101	156
62	79	102	159
63	81	103	161
64	83	104	163
65	85	105	166
66	86	106	168
67	88	107	170
68	90	108	172
69	91	109	175
70	93	110	177
71	95	111	180
72	96	112	182
73	98	113	185
74	100	114	188
75	101	115	191
76	103	116	193
77	105	118	199
78	107	119	201
79	108	120	204
80	110		
81	112		
82	114		
83	116		
84	118		
85	120		
86	123		
87	125		
88	127		
89	129		
90	131		
91	133		
92	136		
93	138		
94	140		
95	142		
96	145		
97	147		
98	149		
99	152		

Exhibit 637-39   CMAI Annual Growth - *Pinus virginiana* (Virginia pine)

Site Index	Ft^3/Ac
40	42
41	45
42	47
43	50
44	52
45	55
46	58
47	60
48	63
49	65
50	68
51	70
52	73
53	75
54	77
55	80
56	82
57	84
58	86
59	88
60	91
61	93
62	95
63	96
64	98
65	100
66	102
67	104
68	105
69	107
70	109
71	110
72	112
73	113
74	114
75	115
76	117
77	118
78	119
79	121

Exhibit 637-40   CMAI Annual Growth - *Platanus occidentalis* (American sycamore)

Site Index	Ft³/Ac	Site Index	Ft³/Ac
60	46	100	123
61	48	101	127
62	49	102	130
63	50	103	133
64	52	104	137
65	53	105	140
66	54	106	143
67	56	107	147
68	57	108	150
69	58	109	153
70	60	110	157
71	62	111	162
72	63	112	167
73	65	113	172
74	67	114	177
75	69	115	183
76	71	116	188
77	73	117	193
78	75	118	198
79	77	119	203
80	78	120	208
81	80		
82	82		
83	84		
84	86		
85	88		
86	90		
87	92		
88	94		
89	96		
90	98		
91	101		
92	103		
93	106		
94	108		
95	111		
96	113		
97	116		
98	118		
99	121		

Exhibit 637-41 CMAI Annual Growth - *Populus L.* (cottonwood)

Site Index	Ft³/Ac	Site Index	Ft³/Ac	Site Index	Ft³/Ac
30	8	69	57	108	150
31	9	70	58	109	153
32	10	71	60	110	156
33	11	72	62	111	159
34	12	73	64	112	162
35	14	74	66	113	165
36	14	75	67	114	168
37	16	76	70	115	172
38	17	77	71	116	174
39	18	78	74	117	177
40	19	79	76	118	180
41	20	80	78	119	183
42	21	81	80	120	186
43	22	82	83	121	189
44	24	83	85	122	193
45	25	84	88	123	196
46	26	85	91	124	200
47	27	86	93	125	204
48	28	87	95	126	207
49	30	88	98	127	210
50	31	89	100	128	213
51	32	90	103	129	216
52	34	91	105	130	219
53	35	92	108		
54	36	93	110		
55	37	94	113		
56	38	95	116		
57	40	96	118		
58	41	97	120		
59	42	98	123		
60	44	99	125		
61	45	100	128		
62	46	101	130		
63	48	102	133		
64	49	103	135		
65	51	104	138		
66	52	105	141		
67	54	106	144		
68	56	107	147		

Exhibit 637-42   CMAI Annual Growth - *Populus grandidentata* (bigtooth aspen) - Lake States

Site Index	Ft³/Ac
40	22
41	24
42	26
43	28
44	30
45	32
46	35
47	37
48	39
49	41
50	43
51	45
52	47
53	49
54	51
55	53
56	56
57	58
58	60
59	62
60	64
61	66
62	68
63	70
64	71
65	73
66	75
67	76
68	78
69	80
70	81
71	82
72	84
73	85
74	86
75	87
76	89
77	90
78	91
79	93

Site Index	Ft³/Ac
80	94

Exhibit 637-43 CMAI Annual Growth - *Populus tremuloides* (quaking aspen) – Lake States Lake States

Site Index	Ft³/Ac
40	22
41	24
42	26
43	28
44	30
45	32
46	35
47	37
48	39
49	41
50	43
51	45
52	47
53	49
54	51
55	53
56	56
57	58
58	60
59	62
60	64
61	66
62	68
63	70
64	71
65	73
66	75
67	76
68	78
69	80
70	81
71	82
72	84
73	85
74	86
75	87
76	89
77	90
78	91

Site Index	Ft³/Ac
79	93
80	94

Exhibit 637-44   CMAI Annual Growth - *Populus tremuloides* (quaking aspen) - Central Rockies

Site Index	Ft³/Ac		Site Index	Ft³/Ac
40	16		80	48
41	17		81	49
42	18		82	50
43	19		83	51
44	20		84	52
45	20		85	53
46	21		86	54
47	22		87	55
48	26		88	56
49	24		89	58
50	25		90	59
51	25			
52	26			
53	27			
54	28			
55	28			
56	29			
57	30			
58	31			
59	31			
60	32			
61	33			
62	34			
63	34			
64	35			
65	36			
66	36			
67	37			
68	38			
69	38			
70	39			
71	40			
72	41			
73	41			
74	42			
75	43			
76	44			
77	45			
78	46			
79	47			

637-107

Exhibit 637-45 CMAI Annual Growth - *Populus tremuloides* (quaking aspen) – Alaska

Site Index	Ft³/Ac
35	2
36	7
37	10
38	14
39	17
40	19
41	21
42	22
43	24
44	26
45	28
46	30
47	33
48	35
49	37
50	39
51	41
52	43
53	45
54	47
55	49
56	50
57	52
58	53
59	55
60	56
61	58
62	60
63	61
64	63
65	65
66	66
67	68
68	70
69	71
70	72
71	74
72	76
73	77
74	78

Site Index	Ft³/Ac
75	80

637-108

Exhibit 637-46   CMAI Annual Growth - *Pseudotsuga menziesii* (Douglas-fir)

Site Index	Ft³/Ac	Site Index	Ft³/Ac	Site Index	Ft³/Ac	Site Index	Ft³/Ac
80	58	120	115	160	170	200	208
81	60	121	116	161	171	201	209
82	61	122	118	162	172	202	210
83	62	123	119	163	173	203	211
84	63	124	121	164	174	204	211
85	64	125	122	165	176	205	212
86	66	126	124	166	177	206	213
87	67	127	125	167	178	207	214
88	68	128	127	168	179	208	214
89	69	129	128	169	180	209	215
90	70	130	129	170	181	210	216
91	72	131	131	171	182		
92	73	132	133	172	183		
93	74	133	134	173	184		
94	75	134	136	174	185		
95	77	135	138	175	186		
96	78	136	139	176	187		
97	79	137	140	177	188		
98	81	138	142	178	189		
99	82	139	144	179	190		
100	84	140	145	180	191		
101	85	141	146	181	192		
102	86	142	148	182	193		
103	88	143	149	183	194		
104	89	144	150	184	194		
105	91	145	152	185	195		
106	92	146	153	186	196		
107	94	147	154	187	197		
108	95	148	156	188	198		
109	97	149	157	189	199		
110	98	150	158	190	200		
111	100	151	159	191	201		
112	101	152	161	192	202		
113	103	153	162	193	202		
114	105	154	163	194	203		
115	106	155	164	195	204		
116	108	156	165	196	205		
117	110	157	167	197	206		
118	111	158	168	198	207		
119	113	159	169	199	208		

Exhibit 637-47   CMAI Annual Growth - *Quercus L.* (upland oaks)

Site Index	Ft³/Ac	Site Index	Ft³/Ac
40	26	79	61
41	27	80	62
42	28	81	62
43	28	82	64
44	29	83	64
45	30	84	65
46	31	85	67
47	32	86	68
48	32	87	68
49	33	88	70
50	34	89	71
51	35	90	71
52	36	91	72
53	37	92	74
54	38	93	74
55	38	94	75
56	39	95	77
57	40	96	77
58	41	97	78
59	42	98	80
60	43	99	81
61	44	100	81
62	45		
63	46		
64	47		
65	47		
66	48		
67	49		
68	50		
69	51		
70	52		
71	53		
72	54		
73	55		
74	56		
75	57		
76	58		
77	59		
78	60		

Exhibit 637-48   CMAI Annual Growth - *Quercus nigra* (water oak)

Site Index	Ft³/Ac
45	27
46	28
47	29
48	31
49	32
50	33
51	35
52	36
53	37
54	39
55	40
56	41
57	43
58	44
59	45
60	47
61	48
62	50
63	51
64	52
65	54
66	55
67	56
68	58
69	59
70	60
71	62
72	63
73	64
74	66
75	67
76	68
77	70
78	71
79	72
80	74
81	75
82	76
83	77
84	79

Site Index	Ft³/Ac
85	80
86	81
87	82
88	83
89	85
90	86
91	87
92	88
93	89
94	91
95	92
96	93
97	94
98	95
99	97
100	98
101	99
102	100
103	101
104	103
105	104
106	105
107	106
108	107
109	109
110	110

Exhibit 637-49   CMAI Annual Growth - *Quercus pagoda* (cherrybark oak)

Site Index	Ft³/Ac	Site Index	Ft³/Ac
70	61	110	189
71	63	111	193
72	66	112	197
73	68	113	201
74	71	114	205
75	73	115	209
76	76	116	213
77	78	117	217
78	81	118	221
79	83	119	225
80	86	120	229
81	89		
82	92		
83	95		
84	98		
85	101		
86	103		
87	106		
88	109		
89	112		
90	115		
91	119		
92	122		
93	126		
94	129		
95	133		
96	137		
97	140		
98	144		
99	147		
100	151		
101	155		
102	159		
103	162		
104	166		
105	170		
106	174		
107	178		
108	181		
109	185		

Exhibit 637-50   CMAI Annual Growth - *Quercus phellos* (willow oak)

Site Index	Ft³/Ac		Site Index	Ft³/Ac
45	27		85	80
46	28		86	81
47	29		87	82
48	31		88	83
49	32		89	85
50	33		90	86
51	35		91	87
52	36		92	88
53	37		93	89
54	39		94	91
55	40		95	92
56	41		96	93
57	43		97	94
58	44		98	95
59	45		99	97
60	47		100	98
61	48		101	99
62	50		102	100
63	51		103	101
64	52		104	103
65	54		105	104
66	55		106	105
67	56		107	106
68	58		108	107
69	59		109	109
70	60		110	110
71	62			
72	63			
73	64			
74	66			
75	67			
76	68			
77	70			
78	71			
79	72			
80	74			
81	75			
82	76			
83	77			
84	79			

Exhibit 637-51   CMAI Annual Growth - *Sequoia sempervirens* (redwood)

Site Index	Ft³/Ac	Site Index	Ft³/Ac	Site Index	Ft³/Ac
122	151	162	246	202	385
123	153	163	249	203	390
124	154	164	282	204	394
125	155	165	288	205	398
126	156	166	288	206	402
127	157	167	261	207	406
128	159	168	264	208	410
129	160	169	267	209	415
130	161	170	270	210	419
131	164	171	273	211	423
132	157	172	276	212	427
133	170	173	279	213	432
134	173	174	282	214	436
135	177	175	285	215	440
136	180	176	288	216	444
137	183	177	291	217	448
138	186	178	294	218	453
139	189	179	297	219	457
140	192	180	300	220	461
141	194	181	306	221	465
142	196	182	307	222	470
143	199	183	311	223	475
144	201	184	315	224	480
145	204	185	319	225	484
146	206	186	323	226	489
147	208	187	326	227	494
148	211	188	330	228	499
149	213	189	334	229	503
150	216	190	338	230	508
151	218	191	342	231	513
152	220	192	346	232	518
153	223	193	350	233	522
154	225	194	353	234	527
155	228	195	357	235	532
156	230	196	361	236	537
157	232	197	365	237	541
158	235	198	369	238	546
159	237	199	373	239	551
160	240	200	377	240	556
161	243	201	381		

Exhibit 637-52   CMAI Annual Growth - *Thuja occidentalis* (eastern arborvitae)

Site Index	Ft³/Ac
25	36
26	37
27	39
28	40
29	41
30	42
31	44
32	46
33	48
34	49
35	51
36	53
37	55
38	56
39	58
40	59
41	61
42	62
43	64
44	66
45	67
46	69
47	71
48	72
49	74
50	75
51	77
52	79
53	80
54	82
55	83

Exhibit 637-53   CMAI Annual Growth - *Tsuga heterophylla* (western hemlock)

Site Index	Ft³/Ac	Site Index	Ft³/Ac	Site Index	Ft³/Ac	Site Index	Ft³/Ac
100	142	128	196	156	248	184	299
101	144	129	198	157	249	185	301
102	145	130	200	158	251	186	303
103	147	131	202	159	252	187	305
104	149	132	204	160	254	188	306
105	151	133	205	161	256	189	308
106	153	134	207	162	258	190	310
107	154	135	209	163	260	191	312
108	156	136	211	164	262	192	314
109	158	137	213	165	264	193	316
110	160	138	214	166	266	194	318
111	162	139	216	167	268	195	320
112	164	140	218	168	270	196	322
113	166	141	220	169	272	197	324
114	168	142	222	170	274	198	326
115	170	143	224	171	276	199	328
116	172	144	226	172	278	200	330
117	174	145	228	173	279	201	332
118	176	146	230	174	281	202	333
119	178	147	232	175	283	203	335
120	180	148	234	176	285	204	336
121	182	149	236	177	287	205	338
122	184	150	238	178	288	206	340
123	186	151	240	179	290	207	341
124	188	152	241	180	292	208	343
125	190	153	243	181	294	209	344
126	192	154	244	182	296	210	346
127	194	155	246	183	297		

Exhibit 637-54   CMAI Annual Growth Reference

Abies balsamea (balsam fir) Picea rubens (red spruce)	Meyer, W.H. 1929. **Yields of second-growth spruce and fir in the northeast.** USDA Technical Bulletin 142.
Acer saccharinum (silver maple)	Brendemuehl, R.H., A.L. McComb, and G.W. Thomson. 1961. **Stand, yield and growth of silver maple in Iowa.** Iowa State University Extension Service F-159.
Betula papyrifera (paper birch) Picea pungens (black spruce) Pinus resinosa (red pine)	Plonski, W. L. 1960. **Normal yield tables for black spruce, jack pine, aspen, white birch, tolerant hardwoods, white pine, and red pine for Ontario.** Ontario Department of Lands and Forests Silvicultural Series Bulleting 2.

Liquidambar styraciflua (sweetgum)	Winters, Robert K. and James G. Osborne. 1935. **Growth and yield of second-growth red gum in fully stocked stands on alluvial lands in the south.** USDA, Forest Service. Southern Forest Experiment Station Occasional Paper 54.
Liriodendron tulipifera (tuliptree)	McCarthy, E. F. 1933. **Yellow poplar characteristics, growth and management.** USDA Technical Bulletin 356.
Pinus banksiana (jack pine) Pinus strobus (eastern white pine) Populus grandidentata (bigtooth aspen) - Lake States Populus tremuloides (quaking aspen) - Lake States	Brown, R. M and S.R. Gervorkiantz 1934. **Volume, yield, and stand tables for tree species in the Lake States.** University of Minnesota Experiment Station Technical Bulletin 39.
Pinus clausa (sand pine)	Schumacher, F.X. and T.S. Coile. 1960. **Growth and yield of natural stands of the southern pines.** T.S. Coile, Inc., Durham, NC.
Pinus echinata (shortleaf pine) Pinus elliottii (slash pine) Pinus palustris (longleaf pine) Pinus taeda (loblolly pine)	United States Department of Agriculture. 1929. **Volume, yield, and stand tables for second-growth southern pines.** USDA Miscellaneous Publication 50. (revised 1976).
Pinus strobus (eastern white pine) - WV, MD, DE, VA	USDA, Forest Service. 1965. **The mensurational characteristics of eastern white pine.** Northeastern Forest Experiment Station Research Paper NE-40.
Pinus strobus (eastern white pine) PA, NJ, NY, and New England	Frothingham, E.H. 1914. **White pine under forest management.** USDA, Forest Service Bulletin 13.
Pinus virginiana (Virginia pine)	Slocum, G. K. and W. D. Miller. 1969. **Reproduction, growth and management on the Hill Demonstration Forest, Durham County, N.C.** North Carolina Agricultural Experiment Station Technical Note 100.
Quercus L. (upland oaks)	Schnur, G. Luther. 1937. **Yield, stand, and volume tables for even-aged upland oak forests.** United States Department of Agriculture Technical Bulletin 560.
Pinus ponderosa (ponderosa pine)	Meyer, Walter H. 1938. **Yield of even-aged stands of ponderosa pine.** USDA Technical Bulletin 630. (revised 1961).
Abies concolor (white fir)	Schumacher, Francis X. 1926.**Yield, stand, and volume tables for white fir in the California pine region.** University of California Agricultural Experiment Station Bulletin 407.
Abies magnifica (California red fir)	Schumacher, Francis X. 1928. **Yield, stand and volume tables for red fir in California.** University of California Agricultural Experiment Station Bulletin 456.
Abies lasiocarpa (subalpine fir) Picea engelmannii (Engelmann's spruce)	Alexander, Robert R. and Carleton B. Edminster. 1980. **Management of spruce-fir in even-aged stands in the Central Rocky Mountains.** USDA, Forest Service. Rocky Mountain Forest and Range Experiment Station Research Paper RM-217.

Pinus contorta (lodgepole pine)	Dahms, Walter G. 1964. **Gross and net yield tables for lodgepole pine.** USDA, Forest Service. Pacific Northwest Forest and Range Experiment Station Research Paper PNW-8.
Juniperus L. (juniper) - Southwest Pinus L. (pinyon pine) - Southwest	Howell, Joseph Jr. 1940. **Pinon [sic] and juniper, a preliminary study of volume, growth and yield.** USDA, Soil Conservation Service. Region 8 Regional Bulletin 71, Forest Series 12.
Alnus rubra (red alder)	Worthington, Norman P., Floyd A. Johnson, George R. Staebler, and William J. Lloyd. 1960. **Normal yield tables for red alder.** USDA, Forest Service. Pacific Northwest Forest and Range Experiment Station Research Paper No 36.

Sequoia sempervirens (redwood)	Lindquist, James L. and Marshall N. Palley. **Empirical yield tables for young-growth redwood.** University of California, Division of Agricultural Sciences. California Agricultural Experiment Station Bulletin 796.
Picea sitchensis (Sitka spruce) Tsuga heterophylla (western hemlock)	Meyer, Walter H. 1937. **Yield of even-aged stands of Sitka spruce and western hemlock.** USDA, Forest Service. Pacific Northwest Forest Experiment Station Technical Bulletin 544.
Larix occidentalis (western larch)	Cummings, L.J. 1937. **Larch-Douglas-fir board foot yield tables.** USDA, Forest Service. Rocky Mountain Forest and Range Experiment Station Applied Forestry Note 78. Meyer, Walter H. 1961. **Yield of even-aged stands of ponderosa pine.** USDA Technical Bulletin 630. (revised 1961).
Pinus monticola (western white pine)	Haig, Irvine T. 1932. **Second-growth yield, stand, and volume tables for the western white pine type.** USDA, Forest Service. Northern Rocky Mountain Forest Experiment Station Technical Bulletin 323.
Juniperus occidentalis (western juniper)	Sauerwein,William J. 1981. **Western juniper site index curves.** USDA, Soil Conservation Service. West Technical Service Center Technical Notes Woodland - No. 14. (adapted from unpublished data in letter by Barrett, James W. and Patrick H. Cochran. USDA, Forest Service Silviculture Lab, August 4, 1981.)
Populus tremuloides (quaking aspen) - Alaska *Betula papyrifera* (paper birch) - Alaska	Gregory, Robert A. and Paul M.Krinard. 1965. **Growth and yield of well-stocked aspen and birch stands in Alaska.** USDA, Forest Service. Northern Forest Experiment Station Research Paper NOR-2.
Populus tremuloides (quaking aspen) - Central Rockies	Baker, F.S. 1925. **Aspen in the Central Rocky Mountain Region.** Unites States Department of Agriculture Bulletin 1291.
Picea glauca (white spruce)	Farr, Wilbur A. 1967. **Growth and yield of well-stocked white spruce stands in Alaska.** USDA, Forest Service. Pacific Southwest Forest and Range Experiment Station Research Note PSW-363.
Pseudotsuga menziesii (Douglas-fir)	McArdle, Richard E., Walter H. Meyer, and Donald Bruce. 1930. **The yield of Douglas fir in the Pacific Northwest.** United States Department of Agriculture Technical Bulletin 201 (revised 1949 and 1961)
Nyssa aquatica (water tupelo) Nyssa biflora (swamp tupelo)	Applequist, M. B. 1959. **Soil-site studies, southern hardwoods.** In: Southern forest soils 8th annual forestry symposium; Baton Rouge, LA: Louisiana State University Press: 49-63.
Thuja occidentalis (eastern arborvitae)	Gevorkiantz, S. R. and William Duerr. 1939. **Volume and yield of northern white cedar in the Lake States.** USDA, Forest Service. Lake States Forest Experiment Station.

Larix laricina (tamarack)	USDA Forest Service. **Yield data, compartment prescription handbook.** Forest Service Handbook 2409.2d, Region 9.
Pinus elliottii densa (South Florida slash pine)	Langdon, O. Gordon. 1961. **Yield of unmanaged slash pine stands in Florida.** USDA, Forest Service. Southeastern Forest Experiment Station Paper 123.
Populus L. (cottonwood) Quercus nigra (water oak) Quercus pagoda (cherrybark oak) Quercus phellos (willow oak)	Baker, James B. and W. M. Broadfoot. 1977. **Site evaluation for eight important southern hardwoods.** USDA, Forest Service. Southern Forest Experiment Station General Technical Report SO-14.

Exhibit 637-55 ESI Forest Plot Field Worksheet

ESI Forest Plot Field Worksheet

August, 2000

Forest Site Plot Number				Location Description
ID	Year	State	County	
10	00	1	5	3 miles N of intersection hwy 34 & RR 2719

Location Data

Cover Type	MLRA Number	State Plane Coordinates / Township, Range, Section			Elevation
		Zone/Secti n	East(X) / Township	North(Y) / Range	
81	133 A	15	27E	11 N	300

Physical Data

Pre ipitation (inches)		Landform	Slope						Position on Slope
Annual	Summer		Percent	Kind	Shape	Micro Relief	Aspect (Azimuth)	Length (feet)	
52	4	LP	2	P	None	None			X

Soil Data

Detailed Profile	Detailed Under-story	Mensur. Info	Soil Series Name	Texture		Terms in-lieu of Texture	Past Eros.	Drain-age Class	Altered Water Rela-tions
				Modi-fier	Type				
Y	Y	N	Bonifay	None	LS	None	1	5	None

Density Data

Understory Abundance				Stand Density		Basal Area (Sq. Ft. per Acre)		Crown Competition Factor
Reprod.	All Woody	Grasses, Forbs	Mosses, Lichens	M or E	Percent	M or E	Sq. Ft.	(Applies only to Lodgepole pine)
1	3	2	1	E	40	M	40	

Tree Data

NSPNS	Crown Class	Tree Origin	Tree Dia.	In. Rad. Last 10 Yr.	Age Estimation				
					Ht. Ring Ct.	No. of Rings	Mea. Pt. Age	Total Age	Total Height
PITA	D	S	18.1	0.6	4.5	52	3	55	99
PITA	D	S	19.2	0.8	4.5	48	3	51	104
PITA	D	S	18.7	0.7	4.5	50	3	53	108
PITA	D	S	18.0	1.3	4.5	52	3	55	102
PITA	D	S	17.1	1.2	4.5	48	3	51	100

Site Index Data

NSPNS	Number of Trees	Curve Number	Average Site Index
PITA	5	690	101

Canopy Cover Data

NSPNS	Canopy %	NSPNS	Canopy %	NSPNS	Canopy %	NSPNS	Canopy %
PIEC2	15						
PITA	85						

Ground Cover Data

NSPNS	Rating	NSPNS	Rating	NSPNS	Rating	NSPNS	Rating
ANDRO2	3	ILGL	3	MYCE	2	SAAL	3
DIVI5	2	LIST2	3	QUNI3	3	SMILA2	3
GESE	3	MAVI5	3	RUBUS	3	VI03	3

Remarks

Recently harvested.

Offshore
--
--
--
--
--
--
--
--
--
--
--
--
3
--
--
--
--
--
--
--
--
--
--
--
--
--
--
--
--
--
--
--
--
--
--
--
--
--

Exhibit 63-57   Determining Site Index for Lodgepole Pine

Lodgepole pine (*Pinus contorta*) commonly grows in overly dense stands, and crowding restricts the rate of height growth. Site index must, therefore, be adjusted for stand density, expressed as Crown Competition Factor (CCF). The procedure is described in the 3-step process below.

Step 1

Determine average height and age of the stand in the conventional manner.

Step 2

Determine the CCF of the stand in which the "site trees" developed. Two alternative methods of obtaining CCF are described:

Method 1 – Estimating CCF from Measurements of Stand Diameters at Breast Height

Establish a density plot for each "site tree." It will be a fixed-radius plot centered on the "site tree" or close enough to the "site tree" to accurately estimate the stand density in which the "site tree" developed. Plot sizes are suggested in Table 1. Plots for any group of four "site trees" should be the same size. Plots may overlap, so some trees may occur in more than one plot and may, therefore, have to be tallied more than once.

Measure the diameters at breast height of all trees on the density plots, and record by 1-inch classes. It is not necessary to keep the data from the four density plots separate, but care must be exercised to locate the "site trees" and density plots that make up any sampling group in a part of the stand where density is representative, and where average spacing between trees and the average diameter of those trees are similar.

Multiply the number of trees in each diameter class by the Maximum Crown Area (MCA) for that diameter class. MCA constants for each diameter class are given in Table 2.

Sum the MCA values and convert to an acre basis. The resulting value is the CCF of the stand. See Table 3 for an example of CCF computations.

Table 1 - Sizes of density plots to use for different maximum heights of site trees

Maximum Height of Site Trees (Feet)	Plot Radius (Feet)	Plot Size (Acre)
Unlimited	52.67	0.2
75	37.25	0.1
53	26.33	0.05
37	18.67	0.025
24	11.75	0.01
17	8.33	0.005
12	5.92	0.0025

Table 2 - MCA values for each diameter class.

Diameter Class	MCA (Per Tree)	Diameter Class	MCA (Per Tree)
1	0.040	11	0.645
2	0.067	12	0.746
3	0.102	13	0.854
4	0.145	14	0.969
5	0.194	15	1.092
6	0.251	16	1.222
7	0.315	17	1.359
8	0.387	18	1.504
9	0.466	19	1.655
10	0.552	20	1.814

Table 3 - Example of Computations for CCF

Diameter Class	MCA (Per Tree)	Number of Trees	MCA (Per Tree)	Sample Computations
7	0.315	23	7.245	
8	0.387	41	15.867	
9	0.466	37	17.242	
10	0.552	47	25.944	Plot Size = 0.8
11	0.645	50	32.250	Total MCA = 159.6
12	0.746	30	22.380	CCF = 159.6/0.8 = 199.5
13	0.854	24	20.496	
14	0.969	12	11.628	
15	1.092	6	6.552	
Total			159.604	

Method 2 – Estimating CCF from Measurements of Basal Area and Average Diameter

Determine average basal area of the stand.

Select the prism with the appropriate Basal Area Factor (BAF) for the density of the stand to be sampled.

Count all trees that are "in" at each sample point. Each sample point should be sufficiently close to each "site tree" to permit a valid estimate of CCF around the "site tree."

Total the trees counted at the sample points, and average to determine the average number for the plot.

Multiply the average per plot by the appropriate BAF. The resulting value is basal area per acre.

Obtain an estimate of average diameter. There are many methods available for determining average diameter. The rule-of-thumb procedure described here is rapid and accurate for the purpose intended.

Walk through the stand sampled, and select what appears to be a tree of average diameter.

637-121

Lay out a transect line across the stand that passes close to each "site tree."

Measure the first 10 trees along this line that appear to the observer to be of average diameter.

Sum those diameters and obtain an arithmetical average. That value is the estimate of average diameter.

Determine the BA/AD Factor by dividing the average basal area per acre by the average diameter.

Determine estimated CCF by substituting the BA/AD Factor in the equation:
CCF = 50.58 + 5.25 (BA/AD)

Where:
CCF =- Crown Competition Factor
BA = Average basal area per acre
AD = Average diameter

Step 3

Select the appropriate site index table from the following list, based on the CCF determined in Step 2.

Exhibit 637 - 58   ESI Windbreak Plot Field Worksheet

ESI Windbreak Plot Field Worksheet

August, 2000

General Data

Windbreak Site Plot Number				P-E Index	Precipitation (inches)		Extra Moisture	Elevation (Feet)	MLRA (Number)	Direc-tion Row 1 Faces	State Plane Coordinates / Township, Range, Section		
ID	Year	State	County		Total Annual	Warm Summer					Zone/ Section	East(X) / Township	North(Y) / Range
13	00	46	041	41	23	17	P	1600	55	W	12	T99N	R65W

Soil Information Data

Series Name	Type	Phase	Soil Descrip-tion	Slope Percent	Slope Shape	Slope Aspect	Soil Trapped by Windbreak		Depth of A Horizon	Restrict-tive Layer (inches to)	Alkaline or Saline	
							Depth (inches)	Texture			Kind	Degree
BOLUGA	SL	D	Y	1	CV	SE	0	—	13	13	N	—

Phase Notes: D - Drained

Maintenance and Response Data

Row Number	Cultivation		Ground Cover	Row to Row Spacing (Feet)	Tree to Tree Spacing (Feet)	Repro-duction	Damage						Rooting Depth (Inches)			Vital Notes (Y or N)
	Past	Now					Insects		Disease		Injury		Many	Common	Few	
							Kind	Deg.	Kind	Deg.	Kind	Deg.				
1	E	P	L	14	5	N	—	—	—	—	—	—	8	31	60	N
2	E	P	L	14	4	N	—	—	—	—	SN	1	8	31	60	N
3	E	P	S	14	8	L	BR	1	—	—	—	—	8	31	60	N
4	E	P	L	14	7	L	—	—	—	—	—	—	8	31	60	N
5	E	P	L	14	8	L	—	—	VR	3	—	—	8	31	60	N
6	E	P	L	14	8	L	—	—	CA	1	—	—	8	31	60	N

ESI Windbreak Plot Field Worksheet
August, 2000

ID	Year	State	County
13	00	46	041

Month/Year Planted __5/89__

Species Performance Data

Row Number	NSPNS	Age in Years	Height in Feet Average of all Trees	Height in Feet Average of Two Trees	Diameter to Whole Inches	Condition	Suppression Left	Deg.	Suppression Right	Deg.	Crown Spread (Feet) Left	Crown Spread (Feet) Right	Height to Live Crown (Feet) Left	Height to Live Crown (Feet) Right	Present Survival in Percent	Soil Within Series Range
1	JUVI	11	10	12	2	G	–	–	2	0	5	5	0	0	24	Y
2	LOTA	11	10	12	–	G	1	0	3	0	5	6	1	1	84	Y
3	FRPE	11	21	24	4	G	2	0	4	1	7	7	4	4	100	Y
4	ULAM	11	29	33	8	G	3	1	5	1	8	8	4	5	100	Y
5	GLTR	11	29	34	5	G	4	1	6	1	7	8	6	6	96	Y
6	ULPA	11	24	27	6	G	5	1	–	–	9	11	5	3	80	Y

Row	Comments	Row	Comments

Tree Measurements Data   Measured R to L

Row/NSPNS	1 JUVI No.	1 JUVI Ht.	1 JUVI Dia	2 LOTA No.	2 LOTA Ht.	2 LOTA Dia.	3 FRPE No.	3 FRPE Ht.	3 FRPE Dia.	4 ULAM No.	4 ULAM Ht.	4 ULAM Dia.	5 GLTR No.	5 GLTR Ht.	5 GLTR Dia.
1	1	10	1	1	10	–	2	21	3	14	26	7	1	25	3
2	3	12	2	2	9	–	3	22	5	15	27	6	2	28	6
3	5	12	2	3	11	–	5	20	5	17	30	7	6	26	4
4	10	6	1	4	10	–	12	21	3⊙	18	28	6	9	26	6
5	11	8	1	6	8	–	15	20	5	19	28	9	11	33	5
6	12	12	2	7	12	–	16	20	5	20	28	9	13	27	4
7				8	11	–	17	21	4	22	29	9	17	35	5
8				9	9	–	18	22	4	23	28	7	18	33	6
9				10	10	–	21	24	5	24	30	8	20	28	6
10				11	11	–	22	23	6	25	35	7	22	27	5

Row/NSPNS	6 ULPU No.	6 ULPU Ht.	6 ULPU Dia	No.	Ht.	Dia.	No.	Ht.	Dia.	No.	Ht.	Dia.	No.	Ht.	Dia.
1	3	22	5⊙												
2	7	21	4⊙												
3	8	22	6⊙												
4	10	24	5												
5	14	26	6⊙												
6	19	27	5⊙												
7	20	25	5⊙												
8	21	26	6⊙												
9	24	25	6												
10	25	26	9												

Page 2

Eastern Forest Cover Types

Code	Name	Code	Name	Code	Name
66	ashe juniper/redberry (Pinchot) juniper	6	jack pine/black spruce[2]	75	shortleaf pine
16	aspen	3	jack pine/paper birch[2]	76	shortleaf pine/oak
11	aspen/paper birch[2]	89	live oak	77	shortleaf pine/Virginia pine[2]
97	Atlantic white-cedar	81	loblolly pine	62	silver maple/American elm
101	baldcypress	82	loblolly pine/hardwood	84	slash pine
102	baldcypress/tupelo	80	loblolly pine/shortleaf pine	85	slash pine/hardwood
5	balsam fir	70	longleaf pine	99	slash pine/swamp tupelo[2]
43	bear oak	71	longleaf pine/scrub oak	111	south Florida slash pine
90	beech/southern magnolia[2]	83	longleaf pine/slash pine	73	southern redcedar
60	beech/sugar maple	106	mangrove	72	southern scrub oak
39	black ash/American elm/red maple	68	mesquite	27	sugar maple
29	black cherry[2]	67	Mohrs ("shin") oak	26	sugar maple/basswood
28	black cherry/maple	14	northern pin oak	25	sugar maple/beech/yellow birch
50	black locust	55	northern red oak	93	sugarberry/American elm/green ash
110	black oak	54	northern red oak/basswood/white ash[2]	91	swamp chestnut oak/cherrybark oak
12	black spruce	56	northern red oak/mockernut hickory/sweetgum[2]	104	sweetbay/swamp tupelo/redbay
10	black spruce/aspen[2]	37	northern white/cedar	92	sweetgum/willow oak
7	black spruce/balsam fir[2]	96	overcup oak/water hickory	87	sweetgum/yellow-poplar
13	black spruce/tamarack	18	paper birch	94	sycamore/sweetgum/American elm
2	black spruce/white spruce[2]	35	paper birch/red spruce/balsam fir	38	tamarack
95	black willow	17	pin cherry	105	tropical hardwoods
42	bur oak	65	pin oak/sweetgum	79	Virginia pine
74	cabbage palmetto	45	pitch pine	78	Virginia pine/oak
86	cabbage palmetto/slash pine[2]	98	pond pine	103	water tupelo/swamp tupelo
44	chestnut oak	100	pondcypress	53	white oak
63	cottonwood	40	post oak/blackjack oak	52	white oak/black oak/northern red oak
23	eastern hemlock	108	red maple	51	white pine/chestnut oak
46	eastern redcedar	15	red pine	22	white pine/hemlock
48	eastern redcedar/hardwoods[2]	32	red spruce	20	white pine/northern red oak/red maple
47	eastern redcedar/pine[2]	33	red spruce/balsam fir	107	white spruce
49	eastern redcedar/pine/hardwoods	34	red spruce/Fraser fir	4	white spruce/balsam fir[2]
21	eastern white pine	31	red spruce/sugar maple/beech	9	white spruce/balsam fir/aspen[2]
19	gray birch/red maple	30	red spruce/yellow birch	36	white spruce/balsam fir/paper birch
109	hawthorn	61	river birch/sycamore	88	willow oak/water oak/diamondleaf oak
24	hemlock/yellow birch	69	sand pine	57	yellow-poplar
1	jack pine	64	sassafras/persimmon	58	yellow-poplar/eastern hemlock
8	jack pine/aspen[2]	41	scarlet oak[2]	59	yellow-poplar/white oak/northern red oak

Exhibit 637-59   Forest Cover Type Codes

Western Forest Cover Types

Code	Name	Code	Name	Code	Name
240	Arizona cypress	206	Engelmann spruce/subalpine fir	207	red fir
217	aspen	213	grand fir	232	redwood
203	balsam poplar	210	interior Douglas-fir	220	Rocky Mountain juniper
222	black cottonwood/willow	237	interior ponderosa pine	243	Sierra Nevada mixed conifer
204	black spruce	247	Jeffrey pine	223	Sitka spruce
254	black spruce/paper birch	248	knobcone pine	224	western hemlock
253	black spruce/white spruce	219	limber pine	225	western hemlock/Sitka spruce
250	blue oak/digger pine	218	lodgepole pine	238	western juniper
216	blue spruce	242	mesquite	212	western larch
209	bristlecone pine	205	mountain hemlock	241	western live oak
236	bur oak	233	Oregon white oak	227	western redcedar/western hemlock
246	California black oak	229	Pacific Douglas-fir	228	western redcedar
255	California coast live oak	245	Pacific ponderosa pine	215	western white pine
256	California mixed subalpine	244	Pacific ponderosa pine/Douglas-fir	211	white fir
249	Canyon live oak	252	paper birch	201	white spruce
226	coastal true fir/hemlock	239	pinyon/juniper	251	white spruce/aspen
235	cottonwood/willow	214	ponderosa pine/larch/Douglas fir[2]	202	white spruce/paper birch
234	Douglas-fir/tanoak/Pacific madrone	231	Port Orford-cedar	208	whitebark pine
230	Douglas-fir/western hemlock	221	red alder		

[1]*Forest Cover Types of the United States and Canada* (F.H. Eyre, Society of American Foresters, 1980)

CONTENTS

PART	PAGE

Part 638.0 – General

Refer to the National Forestry Manual for general
guidance relative to information systems.

Refer to the National Forestry Manual for policy and
guidance relative to database information systems.

Part 638.1 – Database Information Systems

Refer to the National Forestry Manual for policy and
guidance relative to decision support systems.

Refer to the National Forestry Manual for policy and
guidance relative to utility software.

Subject Index

Y

www.ingramcontent.com/pod-product-compliance
Lightning Source LLC
Chambersburg PA
CBHW080330270326
41927CB00014B/3156